Kanban

Workflow Visualized: An Expert's Guide

By

Gary Metcalfe

Gary Metcalfe

Copyright 2018 by Gary Metcalfe - All rights reserved. The following book is reproduced below with the goal of providing information that is as accurate and reliable as possible. Regardless, purchasing this book can be seen as consent to the fact that both the publisher and the author of this book are in no way experts on the topics discussed within and that any recommendations or suggestions that are made herein are for entertainment purposes only. Professionals should be consulted as needed prior to undertaking any of the action endorsed herein.

This declaration is deemed fair and valid by both the American Bar Association and the Committee of Publishers Association and is legally binding throughout the United States.

Furthermore, the transmission, duplication or reproduction of any of the following work including specific information will be considered an illegal act irrespective of if it is done electronically or in print. This extends to creating a secondary or tertiary copy of the work or a recorded copy and is only allowed with express written consent from the Publisher. All additional right reserved.

The information in the following pages is broadly considered to be a truthful and accurate account of facts and as such any inattention, use or misuse of the information in question by the reader will render any resulting actions solely under their purview. There are no scenarios in which the publisher or the original author of this work can be in any fashion deemed liable for any hardship or damages that may befall them after undertaking information described herein.

Additionally, the information in the following pages is intended only for informational purposes and should thus be thought of as universal. As befitting its nature, it is presented without assurance regarding its prolonged validity or interim quality. Trademarks that are mentioned are done without written consent and can in no way be considered an endorsement from the trademark holder.

Table of Contents

Introduction.. - 2 -
Chapter 1: The Origins of Kanban and What Kanban Does........... - 5 -

    The Origins of Kanban.. - 6 -
    The Principles of Kanban... - 7 -
    The Core Properties of Kanban... - 8 -
    Getting Started with Kanban.. - 12 -

Chapter 2: Definitions of Words and Language Involved in Using a Lean Environment.. - 16 -

    Just in Time (JIT)... - 16 -
    Pareto Principle... - 21 -
    Lean Manufacturing.. - 23 -
    Kaizen or Continuous Improvement....................................... - 26 -
    Inventory Management Terms.. - 31 -
    Production Management Terms.. - 41 -

Chapter 3: Preparation for Kanban... - 45 -

    Why a Lean Environment Requires JIT?................................. - 45 -
    The Wastes in Kanban... - 47 -
    The 5's in Kanban.. - 56 -

Chapter 4: The Different Diagrams of a Kanban Process............. - 61 -

    Cumulative Flow Diagram.. - 61 -
    Sankey Diagram.. - 65 -
    Fishbone Diagrams... - 69 -
    Types of Kanban... - 72 -

Chapter 5: Using Kanban in a Supply Chain Environment.......... - 76 -

    What is Supply Chain Management?....................................... - 76 -
    The Principles of SCM.. - 80 -
    How Kanban Can Work in Supply Chain Management.......... - 84 -

Chapter 6: What is the ABC Classification?.................................. - 87 -

    Defining the ABC's.. - 89 -
    The Steps to ABC classification.. - 94 -

Chapter 7: Traditional MRP System vs. Kanban........................ - 100 -

    MRP is the Push..- 106 -
    Kanban is the Pull...- 107 -
    Which of These Options is the Best?......................................- 109 -

Chapter 8: Designing a Kanban System...- 112 -

    What is Value Stream Mapping?..- 113 -
    How to Create a Value Stream Map.......................................- 116 -
    How to Map a Value Stream Using Kanban......................... - 121 -

Chapter 9: Maintaining Kanban... - 127 -

    Kanban Card Maintenance... - 127 -
    Kanban Audits...- 131 -
    Reviewing the Roles and Responsibilities for Kanban Implementation...- 136 -
    Kanban Management for Raw Materials.............................. - 137 -
    Six Rules to Make the Kanban System Effective................. - 140 -

Conclusion.. - 142 -

Introduction

Congratulations on purchasing *Kanban* and thank you for doing so.

The following chapters will discuss everything you need to know in order to get started with some of the more advanced features of the Kanban system. In the previous two books, we discussed the basics of Kanban, about how it works, and more. But now, we are going to delve a little bit deeper into the system in order to get an idea of how it really works, how you can implement it into your own business, and some of the different techniques that you can choose to make it work.

Even though we talked about the Kanban system a bit in our other two books, we will start out this one with a little introduction as well. We will explore some of the beginnings of Kanban and what it is able to do, before moving into some of the words and other languages that is used in the Lean environment. From there, we will move on to an exploration of what you need to prepare for using Kanban, how to recognize and use some of the different diagrams of the Kanban process, and how to use Kanban in the supply chain management and environment to see some amazing results.

From there, we will move on to a discussion about ABC classification and how this can help you to prioritize your

Kanban system and really get some great results in the process. Then we will move on to a discussion about the MRP system and the Kanban system and how these two are similar, how they are different, and the reasons that you may like or dislike each one.

To finish off this guidebook, we will spend a bit of time looking at some of the steps that you need to take in order to design and implement the Kanban system into your business, and how to get your team to fully understand the system so they can get on board. Finally, we will also look at the steps that are necessary to maintain the Kanban system. You can set up the best system in the world using the ideas of Kanban, but if no one continues to use it and it isn't properly maintained, then this system is not going to provide you with the benefits you are looking for.

There are so many benefits of learning about and using the Lean methodology in your own business and to reduce waste and inefficiencies throughout your production process. Kanban can be a visual way that you can implement into your process, in a way that everyone on the team will understand and be able to follow. When you are ready to learn more about the Kanban system and you really want to make it work, make sure to check out this guidebook to help you get started.

There are plenty of books on this subject on the market, so thanks again for choosing this one! Every effort was made to ensure it is full of as much useful information as possible. Please enjoy!

Chapter 1: The Origins of Kanban and What Kanban Does

There are a lot of different processes that are used when it comes to working with a manufacturing company or with software development. One of the methods and techniques that are becoming very popular to use is Kanban. This is a process for software development that helps to provide more efficiency compared to some of the other methods. Kanban underpins the Toyota just in time or JIT production system. Although producing software is a creative activity and it involves some different processes compared to producing a lot of cars, the mechanisms that are needed for managing both of them can still be applied.

When we talk about a process for software development, we can think of it like a pipeline that has some requests that enter in one end, and then these are turned into improved software that will come out on the other end. Inside this pipeline, the business is going to have some process in place that can help get all of this done. The process can range from an informal ad hoc process to one that is very formal and comes in a ton of phases.

For the purpose of this chapter, we are going to look at the more simple process. There will be three main steps that come with looking through Kanban and the pipeline that comes with it including analyzing the requirements, developing the code, and testing out how it all works. Each company can choose exactly how they want to do all of this and the number of steps that you need to take to get it all done.

The Origins of Kanban

Kanban is basically a scheduling system that was developed and used for just in time manufacturing and for lean manufacturing. This system was developed by an industrial engineer from Toyota as a way to improve the efficiency of the manufacturing team. Kanban is just one of the methods that can be used to help achieve JIT. The system will take its name due to the cards that are used in order to track production that occurs within the factory.

Kanban was developed to be used in the manufacturing world, but it has also become an effective tool that can help support the efficiency of many different types of industries and it can help to promote improvement wherever it is needed. The problem areas for the team are going to be highlighted with the help of measuring out the cycle time and the lead time of the full process.

One of the benefits that you will see with the Kanban system is that it can help the team establish an upper limit to the amount of WIP or work in progress so that they don't take on too much and create a bottleneck in the system. In addition, one of the goals of the Kanban system is to make sure that there is never an excess amount of inventory that builds up in any part of the production phase. Limits on the amount of items that can be waiting at supply points can be established and then they can be reduced as different inefficiencies are identified and the team is able to remove them. If the limit is exceeded, this is going to point out that there is some kind of inefficiency that your team needs to focus on removing.

The Principles of Kanban

Kanban is going to be based on three main principles that can make it easier to handle. The three basic principles that come with working in Kanban include:

- Visualize what you need to work on today or the workflow: When you are able to see all of the items in context with the others, it can really inform you of what has been done, what is being worked on, and what still needs to be accomplished.

- Limit the amount of work that is in progress (WIP): This can make it easier for the workflow to be balanced. This ensures that your team isn't going to start and commit to too much work right from the beginning.

- Enhance flow: When an item or a task gets finished, the next highest thing from the backlog is going to be pulled into play and used.

The process of Kanban is going to promote a continuous collaboration. In addition, it asks for an active and ongoing learning and improving system because it will define the best workflow for the team to get things done in the most efficient manner possible.

The Core Properties of Kanban

There are five core properties that come up when you are using the process of Kanban. Knowing what these are, and how they work together, can make a big difference when it comes to how successful your project can be. The five core properties that need to be present in order to see success with Kanban will include:

Visualize your workflow

The team needs to have a good idea of what it takes to get an item from a request from the project owner to completion with a finished project. The goal of using Kanban is to make a change that is positive to ensure optimization of the workflow throughout the whole system. After understanding how this kind of workflow functions in the business right now, you can then aspire to make improvements and adjustments to have it work better. If you try to make changes ahead of time, it can make things difficult.

The best way for a team to visualize their workflow is to use card walls and add in different cards and columns in there. Each column that is on the wall will represent steps that occur in your workflow. Sometimes it can even be beneficial to visualize the work that is coming in and will need to be handled.

The team will have to work together to determine which categories they would like to add to the Kanban board. You can then use different colored Post-It notes to help you address these and give them the categories that seem to work the best.

Limit the WIP

Another thing that you need to work on here is learning how to limit the amount of work in progress. This one implies that the pull system has been implemented on parts or on all of the workflow. The elements that need to be done right away, or are seen as the most important, are going to be the things the team is working on right away. The new work is going to be pulled into the next step as soon as there is some capacity with your WIP limit, but not before. These constrain are going to help you figure out where some of the problem areas are in the flow so you have a better chance of resolving them. Learning how to limit the amount of WIP that you have at a single time can make the team more efficient at getting the work done.

Manage flow

The main reason that a lot of companies are going to implement the Kanban system is because they want to be able to create some positive change in the system. Before you are able to get that change, you must know what change is necessary.

You can figure this out by looking at the way that value is able to flow through the whole system, take an analysis of the areas that are causing problems, then implementing the changes. As you go through and repeat this cycle, you will be able to see what effects the changes had on your system because you will need to know this before moving on. The process has to keep going on, it is one that will never be done, but it ensures that the system works and that your team is being as efficient as possible.

Make the policies for the process explicit

Before the team can do any of its work, the process needs to have a chance to be defined, published, and socialized. Without a good understanding of how the process works and how the work needs to be done, any discussion of the problems that come up will be emotional, subjective, and anecdotal. When all members of the team start to understand what you are doing with this process and what the overall goals are, then it is easier to make smart decisions about any change that can move you to a more positive direction.

One place where this can be important is when you come up with the definition of done. How is the team going to know when they are done with a step, or when they finish one project and it is time to move on to another one? In fact, it is possible to have this done definition for each step of the workflow so that there are requirements before an item can be pulled forward. Some tools, like LeanKit, can let you do this automatically on your electronic board, but you can easily adjust this on a manual board or any other tool you choose to use.

Use models and the scientific method

And finally, Kanban is going to encourage small continuous, evolutionary, and incremental changes that can stay around. In many cases, Kanban is going to suggest that the team uses the scientific approach in order to meet these goals. There are different options that you can choose to make this happen including the Theory of Constraints, the System of Profound Knowledge, and the Lean Economic Model.

Getting Started with Kanban

While we will go into more details about this part as we progress through this guidebook, starting Kanban is not meant to be a difficult process. It is set up as a way to help you develop a product more efficiently than you were able to do in the past. With the help of diagrams and other tools, Kanban can make it easier to help you get your work done, and for everyone on the team to know what works needs to be done. Some of the steps that you can take to get started with the Kanban system include:

1. Map the value stream or your development process: In this, you can ask where do feature ideas come from and who is responsible for these? What are the steps that an idea has to go through to get from an idea to a finished product in the hands of the user?

2. Define the start and then the endpoints for your Kanban system: These should be the place where political control is determined. Don't worry too much during this stage about having a narrow focus because it won't take long before people outside of the span are going to ask to join in as well.

3. Agree: This is the point where everyone needs to agree on the steps that need to be taken and what is going to happen during the process. Some of the things that your team should agree upon before starting will include the following:
    a. The initial limits for WIP and any policies that can be in place in order to change those limits or break them temporarily.

b. The process for prioritizing or selecting the features that needs to be done.
c. The policies for the way you will classify your services. This could be things like fixed delivery date, expedite, or standard. Determine if you need to provide estimates. And when the team has to choose the work they will do, how will they make this selection.
d. How often reviews should be done to ensure the system is still working.

4. Draw up the Kanban board: This can be as simple or as complicated as you would like. All that you really need here is a whiteboard with a marker or some post-it notes. Do not spend too much of your time trying to make this board look perfect. It is going to evolve and change often so keep it simple and allow it to have room to change.

5. Empirically adjust: As more of the project gets done and more requirements come in, there will need to be some adjustments made to the system. Make any of these adjustments to the board as needed.

The idea of Kanban is pretty simple to work with, and it allows a team to visually see what has been done, what is being done right now, and what needs to be done in the future. It is a visual representation, which happens to work the best for most people who are working on a project. They can just glance up and see how the process towards completion is going and feel motivated as things move from Work in Progress over to finished or completed. It is a great way to make a team more efficient and will ensure they get the work done in no time.

Chapter 2: Definitions of Words and Language Involved in Using a Lean Environment

There are a lot of different terms and ideas that come with the Kanban system, especially since it is part of the Lean methodology. Before you are able to successfully use the Kanban system in order to improve efficiencies in the system, you need to have a good understanding of all these different aspects and how they work together. Let's take a look at some of the words and the language that are involved when you want to work with Kanban and the Lean environment to ensure you are ready to implement it in the proper manner.

Just in Time (JIT)

One topic we are going to discuss quite a bit in this guidebook is the idea of Just in Time or JIT. JIT is going to help eliminate the amount of waste and improve quality in a company. In the process of doing this, it will improve the amount of efficiency that occurs throughout the whole company. There are five main ways that JIT is able to improve the efficiency of an organization and these include:

- The JIT method entails sourcing the raw material or item on demand and the work needs to be scheduled based on the order demand for the product. This type of synchronization of supply with production, and then the same for the production to the demand, will help improve the flow of goods and can reduce waste and any need for facilities to store extra product.

- JIT will focus on getting rid of waste in the production process. This can lead to the company redesigning the workspace in order to make a smooth flow of goods or processes throughout. This can minimize how often the product is transported and can eliminate any task that is redundant.

- The traditional process of manufacturing is going to call for what is known as batch manufacturing. This is basically the manufacturing of a particular model in a lot before moving on to a second batch and so on. JIT allows you to manufacture a single component without any delays.

- The JIT system of Kanban can eliminate any of the scope mistakes that are going to show up on the work floor.

- The effort to get rid of waste and get to zero defects can help to cut down the amount of time it takes to manufacture a product, improves productivity, and can improve the quality of the product.

There are a lot of ways that the JIT method is going to help benefit the company. First on the list is the idea of reducing costs. The costs are eliminated in this method because you will get rid of the costs needed for product storage, for inventory, and for raw material. The traditional way of doing this in business is to consider the raw materials or the inventory that you have for a finished good as an asset. But if the inventory is just sitting there, it isn't bringing you any profits and could become a waste. The JIT method is going to turn this idea around and will consider any stock that you have as an opportunity cost and as a waste that should be avoided.

The JIT system is going to source out the raw materials close to the time of manufacturing and then will ship that product out to the customer once it is done, without needing any storage at all. This can lead to a huge amount of savings because the company doesn't need to store either the raw materials or the finished products. Deploying funds that are tied up in raw materials and inventories means extra revenue for the company, which can really help their finances.

In addition, the JIT methods thrust on getting rid of the waste from your production process and working on improving the quality of the product can help out. It will reduce the number of damaged goods that are being sent out and the costs that go to Human Resources to solve a lot of these problems will be cut down as well.

Another way that the JIT method can help out is with improved relationships with the supplier. Because the JIT method depends on the supplier being able to produce raw materials on demand, the benefits of JIT are going to extend to the kind of relationship that you have with your supplier. JIT is going to develop effective communication with the supplier, specifying the exact quantity and the time of delivery of all your supplies. This can help eliminate any miscommunication or misunderstandings that may occur.

The success that you are able to see with JIT will require reliable organizations and suppliers to take the right steps to upgrade the competencies of the supplier and to establish a good relationship with the supplier. When you can build up this good relationship, it ensures that you get the right products each time, and on time, to make customers happy. The exact nature of order and the long-term proactive relationship with your supplier is going to help out even more with saved time because it eliminates the need for any inspections on the goods you receive.

The JIT method is able to help a company adjust their production to the demand they get. If there is a higher demand, then the production will speed up and make more products. If there is less demand, the production can change up that as well to fit with the lower demand. This helps to ensure that you are always able to provide your customers with the products they want, without having a ton of inventory that you had to pay to make and then pay to store until you get enough demand.

A good example of this is in fast food with McDonald's. With the use of JIT principles, McDonald's was able to cater to the order of any type of burger with equal ease, and at the same time, as any other burger they offered. Without the principles of JIT, the workforce would be geared to deliver fast-moving burgers fast, but a peak hour order of a rare item on the menu, or of a special order, would end up causing a lot of issues in the kitchen.

Being able to implement JIT requires some flexibility in the workforce along with a skilled and committed workforce to get it all done. The advantages of this method can even extend to the Human Resources of the company in some of the ways below:

- Adds to more productivity.
- Increases motivation through rotation of the job each person does.
- A more effective use of all of your employees who have multiple skills.
- Investment in training to develop the skills the team already has and to broaden the skill sets.

The big changes that can happen in an organization that has implemented the JIT method answer the question of why use this method. The many benefits of using this method are what help to make it such a successful management strategy to work with overall.

Pareto Principle

The Pareto principle is the idea that 80 percent of the consequences in anything in life are going to come from just 20 percent of the causes. This means that there is going to be a relationship that is unequal between the inputs and the outputs. This is a principle that can remind many companies about the relationship that often comes out of their production line, and sometimes, this principle is known as the 80/20 rule.

The original observation for this kind of principle was linked to a look at the relationship between the population as a whole and wealth. According to the observations that were done, 80 percent of the land that was present in Italy was owned by about 20 percent of the population. Pareto also did a survey of some other countries nearby and found that this same idea was found abroad as well.

For the most part, this principle is going to be an observation about how things in life won't be distributed in the fairest or even way. An example of this could be that the effort of 20 percent of the staff in a company would be able to drive 80 percent of the profits the firm bring in. When we look at time management, either on a personal or a business level, 80 percent of the work output that you do could come from just 20 percent of the time you spend at work.

While Pareto may have been doing this observation to see how wealth was divided in Italy, you can see that there are a ton of other applications that can show up when it comes to these ideas. You can use it in personal time management, in the production process, with inventory, and so much more.

This brings up the question, why is this principle so important? There is actually a very practical reason why you would want to apply the Pareto principle. It basically gives you a window into who to reward or what to fix. If you know that just 20 percent of your products are making up 80 percent of your profits, then it is more likely that you are going to spend the majority of your energy on those products, rather than on the other products and wasting your time. In addition, if you find that 20 percent of the customer base you work with is able to drive 80 percent of the sales that you make, you can change your focus to them and find ways to reward them for being so loyal to you.

As a manufacturing company, you can use the principles that come with this idea to help you divide up your focus the most efficient way possible. You don't want to divide up your time on things that aren't going to bring you profits, for example, if you spend most of your time on things that collectively only bring in 5 percent of the profit for the company, this means you are missing out on the real money makers.

Lean Manufacturing

As we go through this guidebook, you will quickly notice that we talk about the Lean methodology many times. Kanban is one of the processes that are used in the Lean method as a visual way to reduce the amount of waste that is present in the business and to make the team more efficient. Lean manufacturing is going to be a type of methodology that will focus on finding ways to reduce waste within a system of manufacturing, while also finding ways to maximize how productive the team is. This can also be known as Lean or Lean production.

There are a lot of different parts that come with the ideas of Lean, but the five main principles that come with the methodology of Lean manufacturing will include:

1. Identify the value, but look at it from the perspective of the customer: The customer is going to define the value of a product. To be successful, you must first understand what value the customer places on the products and services. This helps you to determine if the product is even something that the customer wants and how much money you can sell that product for and see results.
2. Map the value stream: This principle will require your team to record and analyze the flow of information or the materials that are required in order to produce a product. The goal is to identify all the waste possible. This value stream will take over the entire lifecycle of a

product, starting with the raw materials and going all the way to disposal.

3. Create flow: This one is going to ask the team to eliminate any functional barriers and then identify ways to improve lead time. This will ensure that there is a smooth transition the whole way through to delivery. This flow is very important because it will help your team to eliminate the waste it is experiencing.

4. Establish a good pull system: This means that the team is only going to start out on new work when there is a good demand for it. You will quickly find that the Lean manufacturing method is going to be a pull system, rather than the traditional push system that we have seen in other methods. With the push system, the inventory is going to be determined ahead of time based on a forecast. These forecasts can be inaccurate though and can result in too much inventory. The pull system that is used by the Lean system is going to be the idea that nothing is purchased or made until the demand is there.

5. Try to gain perfection with the idea of Kaizen or continuous improvement: The whole point of working with Lean manufacturing is that the process should

always strive to reach perfection. This entails targeting the root cause of quality issues and then figuring out where all the waste is before eliminating it too.

The ideas of Lean manufacturing are quickly taking over and Kanban is one of the methods that can be used to ensure that Lean is used properly in the business. It is a method that helps put the ideas of Lean into practice, by visually showing what needs to occur during the whole process.

Kaizen or Continuous Improvement

Another term that you should know more about is the idea of Kaizen. This is an approach to creating continuous improvement based on the idea that small and ongoing positive changes in a system or a product will reap big improvements overall. Typically, this is going to be based on a lot of cooperation in the team, and it needs commitment from everyone who is involved. It goes against some of the traditional approaches that want to work with radical changes or those that go from the top down in order to reach the right transformation.

Kaizen is very important when it comes to Lean manufacturing or to the Kanban system. It was developed in the manufacturing industry in order to lower the number of defects that were present, get rid of all the waste, boost how productive the team could be, promote the innovation of those involved, and encourage accountability and purpose for the workers.

Because Kaizen is an idea that requires everyone in the company has to have the right mindset to get it started, there are ten principles which are often the core of the philosophy that have to be addressed. These ten principles of Kaizen will include:

1. Let go of all the previous assumptions the company followed.
2. Everyone needs to be proactive when it comes to solving problems.
3. Don't just go along with the flow or accept things as the status quo.
4. Let go of any perfectionist ideas and instead choose to take an attitude of adaptive and iterative change.
5. As you run into mistakes (and every team does), look for the right solutions at that time.
6. Create a workplace environment where every member of the team feels empowered that they can contribute as well.

7. Don't just take the obvious issue. Work with asking "Why" at least five times to get you down to the root cause.
8. Cull information, as well as opinions, from many sources.
9. Creativity should be used to find small improvements that are lower cost.
10. Never stop with the process of improving.

Kaizen is under the belief that everything in the business can be improved and nothing should be considered a part of the status quo. It is also going to rest on the principle that there should always be a respect for the people of the company as well. Kaizen is going to involve identifying opportunities and issues and creating solutions before rolling them out.

Now, continuous improvement is at the basis of this idea. Because of this, Kaizen comes with a cycle for continuous improvement. Let's take a look at the following steps to see what this improvement cycle is about and to find out the steps that need to be taken in order to execute this process.

1. Get the employees involved: For this to work, everyone needs to be involved. Have the employees help you to find any issues and problems that may come up in the product. Doing this can create buy-in for some change.

2. Find the problem: Using the feedback that you can gather from the employees and then make a list of these potential opportunities and problems. You should make a shortlist of what needs to be done if you find there are a lot of issues.

3. Create a solution: After you have the list of problems and issues that you want to fix, it is time to encourage everyone in the company to come up with creative solutions. All ideas are encouraged during this stage. You can then pick out a solution to go with of all the ideas that come in.

4. Test the solution: Once you pick out the idea that you want to work with, you can then implement it. Everyone in the company will need to be a participant in the rollout. You can create a pilot program or take some other steps to test out these solutions.

5. Analyze the results: At predetermined intervals, make sure to check out the progress of that idea. Determine if this success has been good or not.

6. Standardize: If you find that the results of that solution have been positive, you can then take that solution and adapt it through all parts of the company.
7. Repeat: These seven steps must be repeated on a regular basis. New solutions can be tested out whenever it is appropriate. You may also find at times that new problems or new lists can be tackled depending on where you are in the process.

Inventory Management Terms

Safety stock

The safety stock, or the bugger stock, is going to be the amount of inventory that the company keeps in order to avoid any issues with a shortfall of materials. You want to be careful when coming up with the safety stock. Too little of a stock can result in shortages, but if you have too much of the stock, it can inflate the costs that you incur for inventory.

The good news is that there are a few ways that you can come up with an idea of how much stock you should keep on hand. You can decide the safety stock based on the current demand that you get from the customer, or the demand that you predict is going to happen in the future. You can figure out the safety stock based on the lead time for the product. And you can also use different calculations to figure out the best way to reduce the need of the company for having a safety stock in the first place.

Lead time

Next on the list is lead time. This is going to be the amount of time that passes between the start and the end of a process. This lead time is going to be scrutinized to look for ways to reduce the time that occurs between the conception and the finalization of a project. Companies are going to review all of the data of a process in order to determine if and where inefficiencies occur, and how they can be dealt with.

Reducing the lead time can really help to streamline operations and can improve productivity, which, in the long run, will increase the output and the revenue that a company can bring in. Longer lead times are never a good thing and they will affect, in a negative manner, the manufacturing and sales techniques.

Many times, the lagging lead time is something that the company is able to fix. These lead times are slower because there are wastes in the system, such as unnecessary movements, a process that has not been streamlined, and more. But in other cases, the lead time delay is something that can't be anticipated. For example, if there is a natural disaster, you may need to close down for a few days before sending out a product. If your raw material has a shortage, this can slow you down as well.

Even with these unexpected delays to the lead time, there are some steps that you can take in order to keep the lead time as short as possible. For example, if you have a product that is crucial to production, and a lack of it would make you fall behind and not be able to reach the demands of the customers, then you may have a supplier on backup to provide that material to you, just in case something happens to your primary supplier.

Lot size

The lot size is going to refer to the number of an item that has been delivered on a specific date or the amount that is manufactured in a single production run. This is basically going to refer to the total quantity of a product that is ordered for manufacturing at one time.

If a company doesn't come up with a standard for a lot size, there isn't going to be a standardization of price and value in the product. This can make some problems because customers will never know how much the product is going to be. When the company comes up with the size of a lot, they can work with their suppliers to come up with a good price, one that should remain steady through the long term.

A smaller lot of production is actually preferred when it comes to companies trying to implement the Lean methodology. Inventory and development are going to affect the size of the lot, even though there are other factors in some cases. When the lot size is too big, it ends up creating a lot of waste because you use a lot of parts that aren't necessarily needed, especially if the demand for that product hasn't been able to come in yet.

With a Lean company, a small lot size is preferable. This kind of lot is able to reduce the variability that shows up in the system and can ensure that smooth production happens along the way as well. A smaller lot size can enhance quality, simplifies the process of scheduling, can help to reduce inventory, and ensures that the team is always working on improvement. If you are using Kanban and other Lean methodologies in y our manufacturing, then it is best to work with a smaller lot size so you can get all of these benefits.

Demand through lead time

The amount of demand that is projected for a product from consumers during the lead time from the supplier to the retailer is the lead time demand. Getting this time down as much as possible can be critical to ensuring that the customer is able to get their product in a timely manner. Working with JIT can be a great way to reduce costs in many ways, but if the customer has to wait too long to get the product through this system, it can lead to a loss in customer satisfaction, and a loss in overall sales.

A failure to estimate the right lead time demand for a product can result in some different issues including shortages in inventory. Letting this occur too often and for too long of a time can cause dissatisfied customers. Your customer may be willing to wait a little bit to get the product, especially during peak seasons. But if his lead time takes too long, they will decide not to purchase from you and will choose to go to one of your competitors.

With the JIT method, you will work to only make the amount of product that is needed, and the goal will be to keep the lead time as low as possible. But, if there is a product that is in too high of demand or a season that the demand is going to be too high, and you can't limit the lead time, it may be acceptable to store a small quantity of the item as a buffer to make sure that the customer is satisfied.

For example, if your company makes a toy that has been really popular, and it has been selling out through other parts of the year, there may be times when you want to keep a small stock of the product. If the company offers a discount on the product at Christmas, you may need this buffer because it would be too hard to create a lead time that can keep up. Then, when the holiday is over, you can cut out the inventory and bring your costs back down.

It is important to know what your lead time is, and determine if it is enough to keep up with the demand that you get from the customer. You need to make the lead time as short as possible so that you are able to keep up with the demand from your customers, especially when you rely on the JIT method. Make sure to figure out this lead time, determine if it can actually keep up and if it can't, figure out ways to reduce the lead time to more acceptable levels.

ABC analysis

The next thing that we need to look at is known as an ABC analysis. With this kind of analysis, you are going to take a subject and divide it up into three categories of A, B, and C.

The first category is A. This is going to be the category that will represent the most valuable products or customers that you have. These are all the products that have the biggest impact on your profits, without taking up too much of the resources that you have. This kind of category is going to be pretty small because it should only contain your biggest money makers and none of the other products you produce.

An example of this may be a company that produces software. They may produce several different kinds of software, but they find that one niche in particular can be sold at a higher price compared to the others, even though it takes the same or fewer resources to make. This is why that particular niche ends up accounting for 60 percent of the overall revenue. Even though the company sells fewer of these products, it is a niche that has a higher percentage because it can be sold for a higher price. This kind of product should be placed in category A.

Next on the list is category B. This is going to represent the products or the customers that are middle of the road. Some businesses are going to take a look at this category in the wrong manner. They assume that this is the group that does contribute to their bottom line but still isn't big enough to get a ton of attention. But the way that you should look at category B is the products that have a lot of potentials. The products that you place into category B are the ones that could be encouraged and turned into items that fit into category A.

And finally, there is category C. This one is going to be focused on all of the smaller transactions that are going to be very essential for profit, but which, when they are taken individually, won't contribute much value to your company or its bottom line. This is the category where you will place most of the products that you sell. It is also the category where you want to make the sales as automatic as possible because this can drive your overhead costs down and increases how profitable they are.

The main reason that you want to use an ABC analysis is to make it easier to deal with the large and complex data you have, just by breaking them into three segments. These segments can be used in order to prioritize the data within the area that you want to use them. Once you break down all of your data into these three segments, it is easier to get a good focus on the data and then learn how to use it in the most meaningful way. Breaking them down can make the different issues that come in the data more obvious and can ensure that you prioritize the right things.

Bill of material

A bill of materials is another thing that you can keep track of the inventory that you want to use in your company. The BOM is going to be an extensive list of raw materials that are required in order to construct, manufacture, or repair the service or product that you have available. This is usually going to appear in a more hierarchical format, with the higher levels displaying the product when it is done, and then the bottom levels are going to show all of the materials and components that are needed to make the product.

There are different types of bills that you can work with and it will depend on the business needs and the projected use of them. For example, a manufacturing BOM is going to be important when it comes to an ERP and MRP of the company. A BOM explosion is going to display an assembly at the highest level that is then broken down into the individual components and parts near the end and links to show how it all comes together.

Let's look at an example of this one. A computer is going to be at the top and then it is going to be exploded down into the processors, the memory panels, the computer chips, and the hard drives. Each of the processors would then be exploded down into the register, control unit, and arithmetic unit. This would keep on happening until the product is down to the smallest parts.

Every line that is found in the BOM is going to include a lot of information. You may find information such as the features of the product, the weight, the length, the size, the unit of measure, quantity, description, part revision, part name and number, and the product code to name a few. This list can be important and even necessary when it comes to ordering replacement parts in your company, and it can reduce the possible issues if any repairs of the product are required. It can also help plan for acquisition orders and can limit the amount of errors that are possible.

Backflush

Backflush costing is often a product costing system that is used in the JIT method. This kind of costing is going to delay the process of costing until all of the goods are produced and done. The costs will then be flushed back to the end of the production run and assigned to the goods that you worked on. This can be a great way to do it because you don't have to worry about detailed cost tracking during production. Since it is a method of accounting, it is pretty common to see backflush costing known as backflush accounting.

By taking the time to eliminate work in process accounts, the idea of backflush costing is going to make the accounting process easier to work with. However, there are some issues because the simplification and other deviations that come with backflush costing compared to other costing systems is that the former doesn't always conform well to the generally accepted accounting principles. In addition, this kind of costing runs into issues of finding a good audit trail.

The companies that decide to do this kind of costing will meet the following conditions:

- Management will seek an accounting system that is simple and that doesn't have a lot of details when they track direct material and direct labor costs.

- Every product they work with has a set of standard costs.

- Material inventory levels are either constant or low. When the inventories are low, the bulk of manufacturing costs will flow into costs of goods sold and it won't be deferred as inventory cost. This kind of costing is particularly attractive to companies that have low inventories resulting from JIT inventory of manufacturing strategies.

Theoretically, backflush costing is an elegant solution to the many complexities that can come when assigning costs to inventory and production, but sometimes a company will have trouble when it comes to implementing it. It can be hard for you to use it on extended production processes, and it can take too long for the inventory records to get their reduction after the product is completed. It is often not going to be used for the fabrication of customized products since this requires you to create a new bill of materials for each of the items that you want to create and manufacture.

Production Management Terms

Takt time

Knowing all about takt time can make it easier for a business to estimate the service delivery process and the outcome of a software or process. It can be a good way to achieve a consistent and continuous flow of production, find ways to eliminate the waste of overproduction simply by having the team only produce what the actual demand of the customer is, and developing a standardized instruction for work. Takt time can even enable you and the team to set real targets for production, ones that are able to show the rest of the team where they need to focus all of their output efforts.

Figuring out the Takt time doesn't have to be too difficult when it comes to using it in the production cycle. To figure this out, you just divide your available time by the rate of demand that you get from the customer. Let's take some time to explore this a bit more and see exactly how the Takt time works and how it can assist you.

1. Calculate the demand from the customer. This can also be seen as what your customer will typically want during a certain time period such as each day, week, or month.

2. Calculate out the available time you have to work on the project. You can exclude any breaks, weekends, and

meeting times since it isn't likely that you will work during those times.

3. Calculate out the Takt time. This is the available time divided by the customer demand that we got in the previous two steps.

4. From there, you can compare the cycle time that you already have for the production cycle against the Takt time. You can choose any chart that you would like to do this, but often a bar chart is the one that is most recommended.

5. From there, you can draw your own value stream mapping. Use this map in order to provide the Takt time for each step in the activity.

Takt time can be so important when it comes to how successful you are going to be with a product and getting it out to the customer once the demand comes in. Takt time is the place of production that aligns your production with the demand of the customer. So, the Takt time is going to vary based on how much demand there is for that product. To make it easier, the Takt time is going to be how fast you need to manufacture a product in order to fill the demand from the customer.

Let's look at an example of this. If you have a customer demand of 100 light bulbs a day, the Takt time is going to be 8/100. We get the eight from the assumption that you are at work for 9 hours during the day with an hour for lunch, leaving you with eight hours to get the work done. This means that you need to be able to get a bulb done every 4.8 minutes to keep up with the demand.

Below is an example of how the Takt time would look when compared to the production cycle for getting the product done and out to the customer.

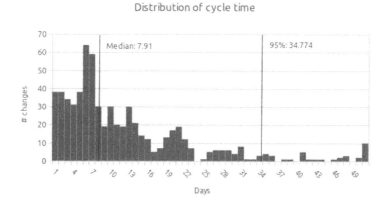

Chapter 3: Preparation for Kanban

At this point, it is time to prepare yourself and the team to use the Kanban system. In order to use it properly, it is important that every member on the team understands how the Lean environment relies on JIT, how to recognize and eliminate the most common wastes that need to be taken care of in the Kanban system, and the 5's that are critical as well. Let's take a look at all of these to gain a better understanding of how they all work.

Why a Lean Environment Requires JIT?

When you are working in a Lean environment, you will find that the methods used in JIT are going to be very important. Just in Time production is going to help reduce the waste of a company, which can be very important when it comes to the Lean methodology. Since a company can use JIT methods in order to only produce the products that are needed at that time, it helps to reduce a lot of costs that come with making too much of the product and then having to store it until the customer places an order.

Lean is all about efficiencies as well. A company that is using the ideas that come with Lean is working to decrease the amount of waste that they are dealing with, to ensure that the process is as smooth and efficient as possible. While it is possible to do this and implement some of the ideas that come with JIT, adding these principles into the Lean methodology can make it even more powerful for a company.

In addition to using the ideas that come with Lean, the JIT can ensure that the production is only making the exact number of products that are needed. You will have less unnecessary movement, more efficiency with the work that customers are doing, and less storage space and wasted money in the process. For Lean to be effective and actually work, it is critical that a company also implement the ideas of JIT along as well.

JIT manufacturing is so important when it comes to working in a Lean environment. It ensures that the wastes are kept low during this process. Since you are making the product pretty much on demand, without worrying about having a stock or inventory to maintain during this process, you are able to easily follow all of the methodologies of Lean. JIT manufacturing helps to make this happen easier.

The Wastes in Kanban

Since Kanban is a Lean method, it makes sense that part of the reason to use the Kanban system is to help eliminate the amount of waste that is present in a system. This method works a little bit differently than you will find with some of the other Lean methods because it focuses on using visualization so that members of the team are actually able to see some of the work they are doing, the work that has been completed, and the work that is still waiting to be added to the WIP.

Even though the Kanban system is a bit different than the others, it can be so effective and it is still able to handle wastes in the same way as the other systems. Kanban should be implemented as a way to deal with these wastes and make sure that they are reduced and eliminated when possible. The seven wastes that are recognized by the Kanban system and will be reduced or eliminated will include the following:

Overproduction

The issue of overproduction can occur when a company is producing way more of its inventory than the customer is demanding. This is going to include both the production of products and some of the components that make the product.

These products are produced even though there are no orders or when there are more produced than there are orders for.

This can be a big waste for the company and it is often seen as one of the worst because it has a way of multiplying some of the other kinds of wastes that the company has to deal with. For example, overproduction can increase the rate of rework, the amount of inventory, the amount of processing, waiting, and lots of unneeded transportation and motion.

This is a waste that needs to be limited as much as possible. It causes a bunch of other wastes in the company, and since there are no customers to purchase the product, it is hard to regain profits from it. Figuring out the best way to limit overproduction and only make the products that customers actually want is critical when using Kanban.

Inventory

Next on the list is the waste of inventory. Inventory is going to refer to the quantity of items the company has in stock, which will be required in order to make a product. These goods can sometimes cause the business some costs. When the inventory is not used, they are going to take up a lot of warehouse space that is seen as valuable. And in some cases, it is possible for them to become obsolete.

In some cases, this inventory may require raw materials. If the inventory isn't being sold or used to make a new product, you are wasting it because it can't be used for some more important goods. Organizations that wish to be more competitive are learning how to control their inventory. This can ensure that the money that was put into unused inventory isn't wasted on items that are not wanted at the time.

The nice thing about the Kanban system is that you are able to reduce the amount of inventory that comes with the production cycle. In traditional manufacturing processes, there is a lot of waste of inventory. The business will try to forecast the amount of demand there is going to be for the product and then they will stockpile it up to be ready for the demand. But this adds the costs of manufacturing. You have to produce more of the product ahead of time and then you have to store it. And if the forecasts were wrong, you then have extra products that you have to try and sell later on and store it even longer.

With the Kanban system, you don't have to worry about this issue. It uses JIT manufacturing so you only produce the product that the demand needed. When a customer places an order, the order will be sent out and that is when the product is produced. This helps to reduce the amount of waste and money lost, compared to some of the other methods of manufacturing that are out there in other businesses.

Defects

Rework is going to be needed any time that a component or a whole product is defective or has a lot of damages on them. Defects are going to be caused when there are bad processes in manufacturing. Sometimes these are going to be done by an error in the machine and sometimes the error is human.

This rework can be bad for the product and the company because it is going to take extra time to get the product fixed and ready for market. This increases the amount of costs to make the product. And in some instances, there are no methods for fixing the product and they will have to be thrown out, which can make it a complete waste of money.

When you work on the Lean methodology and Kanban, it is important to reduce the number of defects that are shown in each product. These defects are going to slow your business down. They may not seem like such a big deal, but often they have a lot of far-reaching effects on the system.

To start with, a defect that isn't caught quickly can end up causing a lot of issues as more and more products have that defect. This means that a lower quality product is going out to the customer. The customer will not be happy, the profits will go down, and your business will have to go back through and find ways to fix this issue as soon as possible.

Even if the defect is found quickly, it can still halt up production in the company. The team has to figure out what the defect is, how far it has gotten, and how to fix it. There may be some changes that are needed in the system to fix this defect as well. And other parts of the manufacturing process will need to be adjusted in order to give that part of the process time to fix the issue. It is much better to reduce the number of defects in the system so that the waste doesn't happen.

Waiting times

Each step in the process is going to be dependent on its upstream and downstream stage processes. If one part of the stream isn't keeping up or ends up falling behind for some reason, then this is going to affect the rest of the system and could cause everyone to fall behind, or at least cause them to just wait around and not get anything done.

If materials, information, equipment, or employees are delayed at all, then it can be a lot of waste. The production time is going to be wasted because these processes are just waiting around to finish their jobs instead of moving on to the next step or processing more stuff in the meantime. In addition, the cost of production will increase, it takes more time and more wages to make a product when the process delays or stalls.

As you go through your Kanban system, look for ways that you can reduce waiting times. The more there are waiting times, the more waste there is. Employees are not working on producing a product. They are still getting paid, but they will have to wait until the next step of the process before they see any results. This adds to the costs of manufacturing that most companies will want to avoid.

If you are creating the Kanban system, see if there are any unnecessary waiting times that are going on in the system and then work to reduce these as much as possible. Try to streamline the process as much as possible so everyone is always working on something and no one is waiting during the process.

Transportation that is not necessary

This type of waste is going to include any of the movement that happens with components, products, and information from one area to another that is unnecessary. This kind of transportation is going to occur along with product damage, lost parts and systems, and unnecessary movement. The more times that you move a product around or any of its components, you are wasting time and money at the same time.

Each movement is going to take time, even if it is just a few seconds more than what was being done originally. It takes the time of your employees to move these items many times and even to figure out where the items are going to go next more times. It takes resources to actually move those products. Finding ways to cut down on the unnecessary transportation can speed up production times, can save money, and more.

The good news is that since you are working on controlling your inventory better and only making the product when the demand is there, such as after a customer places an order, there should already be a reduction in the amount of transportation that is needed. You can cut out the movement to a warehouse to store the inventory for example. As you are looking through your value stream map (we will touch more on this in a bit), check to see when transportation is needed for the product and then determine if that is actually needed, or if you are able to cut it out to reduce this waste.

Unnecessary motion

Any of the unnecessary movements that you want to avoid will often occur when an employee is moving around their workspace and ends up wasting time and effort at the same time. If the employee has to take more steps or move around more than the minimum just to get the work done, then this is going to be a waste of motion.

There are all kinds of unnecessary motion that occur when it comes to poor working standard practices. These can also happen if the work area isn't laid out very well or when the process design is not optimal. Figuring out which of these issues is the one causing the unnecessary motion in your business can be critical to reducing this waste and the Kanban system can help with it.

There are a number of ways that you can make sure to remove any of the motion in the process that is unnecessary. If the process includes management approval before moving onto each step, this is some movement that can be reduced as many times the teams know how to handle the quality of the product, or can be trained how to deal with this. If two teams or departments that need to work together are on opposite sides of the building, it may be time to move them closer together to avoid movement.

If a team or an employee has to walk to different parts of the production line or different parts of the building in order to finish one process, it is time to move these closer. In fact, each team member should be able to have all the items they need within arms reach while they are working. This can help to reduce waiting time and unnecessary movement and can increase the amount of efficiency that you get from that employee.

Over-processing

And the final Kanban waste is going to be the idea of over-processing. Over-processing is going to include any extra steps that are used during the process which need to be taken. If you have a poor layout or you add in some extra steps to the process and require your employees to do them each and every time, this can be an example of over-processing.

In addition, over-processing can mean producing products that are much higher quality than is actually required. This could be due to using the wrong type of equipment, errors in the reworking process, poor design process, or lots of bad communication. Taking the time to check into what the customers really want, rather than just making assumptions, can help you to avoid the waste of over-processing.

Not using the ideas of your employees

This one is not seen as a waste when it comes to Kanban, but it is still an important thing to talk about here. This one is going to include a waste of learning opportunities, improvements, ideas, skills, and time because you don't take the opinions of your employees into account. This is going to mean that employees aren't going to take part in the design or the manufacturing process.

The only way that you are going to be able to get new ideas to develop, the ideas that are able to eliminate and even avoid the other wastes that we talked about, is to listen to the ideas of your employees. This can help you to improve the processes that you are already doing and can increase the amount of creativity and knowledge that becomes a part of the whole process. In addition, it will help you to increase employee satisfaction in the work that you are doing on a daily basis.

The 5's in Kanban

Another topic that we need to look at when it comes to Kanban is the idea of the 5s. When you work on a Kanban system, you may wonder if there is really any value in the way that you go about the system. The team may feel that this is just another way that they have to manage their work, but they aren't sure about all of the benefits. But when you base your system on the 5s, you will find that the system can be even more effective than before and these may even help you keep yourself on track. The 5s that you should know about the Kanban system include:

Seiri

This is the idea of cleaning out or getting rid of any task that the team is doing or did do before they added in Kanban that helps keep track of all the tasks and work that needed to be done. If you are going to use more than one method in order to keep track of all those tasks that you need to get done, you are taking on too much and it won't be long before you even need another tool to manage those original ones. The point of using Kanban is to make things simple, to ensure that you don't have to remember things, and to keep all of your tasks in one place. The idea of this is ruined if you are searching around too many organizational tools and tracking systems to do this.

Seiton

This one will ask that you keep the Kanban organized so that you can use this system as efficiently as possible. It doesn't matter which use of Kanban you choose to go with, but make sure that everything you have is as accessible as possible. If you are using a whiteboard to make this happen, make sure that you are stocked up on post-it notes in different colors, lots of pens, and even a filer so you can store the tasks that you finish in case you need them later.

No matter how you want to do the system, you want it organized in a way that you won't have to go through and search for the things you wish to work on later. The tools shouldn't be kept in a position that is uncomfortable or one that is too far away. Keep it all together, make sure that it is usable, and make it as organized as possible.

Seiso

Next on the list is the idea of Seiso. This one means to shine items or clean things. This one is a bit more complicated than just wiping off the whiteboard on occasion, although that can be a good thing to focus on to keep things clean and organized. The bigger part of this is that you want to make sure that the Kanban method that you are using, whether it is a whiteboard or online, is tidy and kept in good shape.

Before you head out at the end of the day, stop and take a look at the Kanban and see how it is looking. Make sure that it is representative of the work that you are doing, and mark tasks as complete if they are done or no longer needed. If you want to make notes on tasks to do later, then take some time to do that now. Rearrange the work that is left, reorder it, and get the board to look good. These will only take a few minutes of effort at the end of the day, but they can do wonders when it comes to working on your system later on.

Seiketsu

This step is the one of standardizing things. Your team needs to define for themselves a method for the Kanban and then they need to stick with it. If you choose to use different shapes or different colors for the various tasks that need to be done, then make sure that you are consistent with this. If you like to categorize the different tasks, then make sure that you don't use a different category during the process because you can't find the right color or because you were too lazy to use it. You, and everyone else on the team, need to make sure that you are consistent with the work that you are doing and that you can stick with it along the way.

Shitsuke

This means that the team needs to be disciplined enough in order to sustain the order of things as everyone has decided on. Pay attention to the four steps that we listed above regularly. Make sure that you stick with the system that the team defined, that everything is clean and tidy and that you have all the items that you need to keep using this system. And the most important thing is to commit to what you would like to achieve and don't give up on it.

Without the right amount of discipline in place, the method will start to deteriorate over time and the team will go back to their old ways. But if you use discipline and stick with your hard work, you will see that the team is more efficient, and there is a huge improvement in the workflow for everyone.

Chapter 4: The Different Diagrams of a Kanban Process

There are a few different diagrams and methods that you are able to use when it comes to working on the Kanban system. These diagrams are nice because they allow you to see what is going on with a project. If it is used properly, it can help everyone know what is going on with the project every step of the way and it can even help you to spot problems that may come up in the system. Some of the most common diagrams that you may want to implement when using Kanban method includes:

Cumulative Flow Diagram

A cumulative flow diagram is going to be one of the most used tools when it comes to the Kanban method. It will allow your team to have a way to visualize all of their efforts and the progress that a project is making. When there is some issue or another impediment that will occur in the process, the cumulative flow diagram is the area where you will notice this first. Instead of the graph sticking to a smooth and gentle rise, there is going to be a bump or a sudden change that shows up in the graph. When you want to be able to predict what problems are going to occur in the future, then the cumulative flow diagram is the right one to go with.

The first thing to look at here is what exactly the CFD is going to show. This diagram is going to show the way that tasks are done in a company and the way that the work is distributed throughout the stages of the process. The graph is going to be built up from different band colors, with each task being gathered in different columns. To keep things organized, there needs to be just one color for each column. The color of the band is going to let you know how many tasks sit in that stage of the process at any time that you choose to look at it.

So, how is the CFD supposed to look? Ideally, with a diagram that is working the right way, you want to see that the bands are evenly rising as one. The bands should stay pretty much even, except for the band that is in charge of completed tasks. That one needs to have a continuous motion of getting taller because you want the number of tasks that the team completes to continue growing.

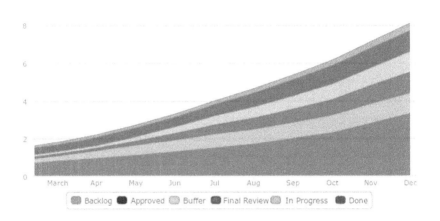

Now that you know what the diagram should look like, the next question is what you need to look out for when you are using this kind of diagram. First, you never want to see any of the bands rise suddenly. This is going to point out an issue with the production that you need to work on. From a successive accumulation of tasks that show up in one band or another, it is possible to see the issues with bottlenecks before they happen, and you can then go through and make an effort in order to prevent these from happening.

Another thing to watch out for with this kind of diagram is that a band that has tasks that are in progress starts to widen out vertically. This means that the amount of tasks that the team is trying to handle at that time is too large and the project as a whole is going to be delayed unless something is fixed in the process.

To create one of these diagrams, you only really need three basic things. These include the backlog, a column for the tasks that are in progress, and then a done section. When you use these three categories, it makes it easier for the team to see information that is present in the diagram easily. You can choose to divide it up into as many categories as are needed to complete your project and make sure that the team is easily able to see what is going on through the project.

The CFD is a good tool to use to help you visualize things. As long as everything stays even and keeps going in the same direction, you know that production is going well and everything is on track. But if you are looking at the chart and you start to notice that the bands spread out or they aren't even any longer, then it is a sign of the process going wrong and you need to make some changes to eliminate the issue or at least reduce them.

Sankey Diagram

Diagrams that are considered Sankey are going to be used in order to provide a team with an analysis of flows that is visual. While it may do the best when showing the flow of materials when we are looking at a distribution system, Sankey diagrams can be adapted to provide this kind of analysis no matter what kind of work is done. This means that a Sankey diagram is going to be effective no matter what kind of workflow you are trying to manage, especially if you are using Kanban.

In this kind of diagram, there is going to be a set of nodes that will be used for a 2-D array and then these two nodes are going to be linked together with some lines. These lines, which are known as links, will represent the flow that comes from one node to the next. The width that comes to each of the links will be able to reflect the volume of the flow between the nodes that are linked.

The vertical axis will show the categories that come with the nodes, with the source categories found on the left, and then the destination categories are going to be arrayed as you need by the structure of the network that you are making. The limits on this are only as far as your imagination goes, and as long as you can clearly convey the information, you can keep working with the Sankey diagram.

Now, the next question to ask here is how you can use this kind of diagram to help you manage the workflow that is done in Kanban. First, the Cumulative flow diagram and a Sankey diagram are going to be very similar in the idea that the width of each part of the diagram is going to have some different changes throughout time to show the absolute volume of flow. However, the biggest difference between the two is that the CFD is going to include one node with a single flow, but the Sankey can have many.

The first way that you can use the Sankey diagram is to make it easier to show the volume of items that need to be worked on per customer for those items. Looking at the graph below, you can see that the chart is going to start showing the quantities for each work item, including service request, problem, change, and incident. It can then show the relative volumes by client. And then, when you look at each client, it will show you the quantity of work items that have been requested. This may be a lot of information, but the Sankey diagram below will show how this is done and will ensure that the team is able to get through and see what is going on.

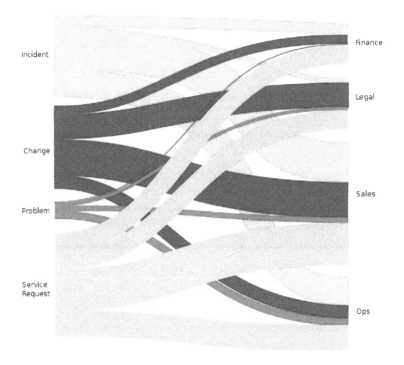

Now, you may see that your impression of this, when based on the thickness of the line, isn't completely precise. It is easy to add in the work items that you need to each category and each of the flow lines, and you can also add in more customers to the count if needed. You can also choose to add in some other statistics as needed, and you may find that adding in some such as the standard deviations, the mean value, and the percentage of the whole, can be placed into this diagram as well.

Another option that you can do is the classes of service versus customer. In the diagram that we will look at below, you will see that the analysis is a bit more complex and there will be three nodes for each flow. It will show us the relative quantity by class of service based on each of the work items. When you look at the diagram, you will see that there are three classes here including the fixed, standard, and intangible date that have fewer links going out compared to the amount that is coming in.

When you work with this diagram type, the inputs to a class of service node aren't allowed to be smaller than the outputs that you get. A more formally complete diagram would also have an abandoned node parallel to the various customers.

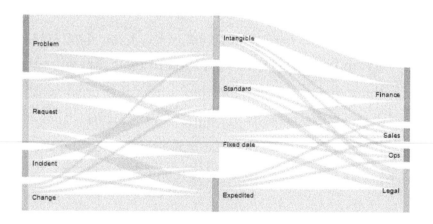

And finally, we will look at team handoffs. The number of times that the product is handed off from one team to another is also a factor to consider when you are working on lead time. A Sankey diagram is also a good way to visualize the flow of work from each team in the company. And then you can use it to see exactly where these handoffs occur, and where the product may be at any given time. This diagram is also used as a way to get a visualization of how many requests have been left behind or abandoned by the teams.

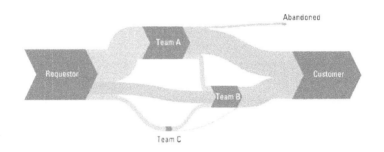

These are just a few of the ways that you can use the Sankey diagram. But now it is time to create one of these diagrams to work for your options. There are different tools that you can use for these diagram types, but they are hard to create with a spreadsheet and some software options won't work either. Google has a source code that can make this easier and the code is pretty easy to choose as well if you do.

Fishbone Diagrams

The fishbone diagram is often known as the cause and effect diagram. This kind of diagram is going to help organize possible causes into a visual format, one that is quickly understood. Making a fishbone diagram is pretty easy to work with. The diagram that we have below is an example of a fishbone diagram which a customer service team made in order to address an issue they were having with customer call times that were taking up too much time.

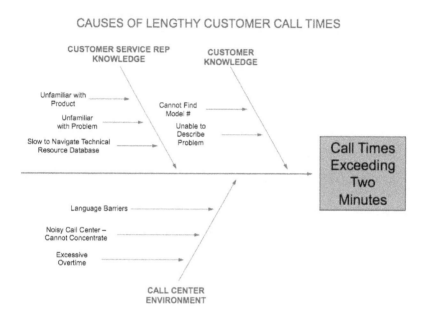

Using this diagram, the team was able to follow a few steps in order to create that diagram and then see results at fixing the issue. The steps that were used in order to create the fishbone diagram that we are using include:

1. State the problem: The statement of the problem is going to be placed as the head of the fish. For this case, the statement is going to be the fact that the call time is going over two minutes. This statement is a good

example because it is specific and it will help the team stay focused.

2. Document the causes that are possible: Before you fill in the skeleton of the diagram, the team needs to get together and brainstorm all of the causes that are making the team have really long call times. Then they will take all of those causes and group them into three categories. In this case, they were grouped into the call center environment, customer knowledge, and the knowledge of the customer service representatives. Your fishbone diagram can certainly be longer if needed.

3. Complete the rest of the diagram: For this example, there were eight potential causes that were assigned to those three groups that we talked about before. The causes were going to include things like creating an environment that is better for the call center, teaching customers more about the product, and better training of call center employees.

After taking the time to finish this kind of diagram, the team was then able to collect the data for two weeks using a check sheet. The team then went through and tallied up all of the reasons for their calls going above two minutes. The team can then go through, figure out the reasons that the call times are bad, and then improve these call times by quite a bit.

Types of Kanban

If you are looking for ways to implement a Kanban system or you would like to improve the way that yours already works, then you must make sure that the proper Kanban cards are available for your business. These types of cards are going to be the main tool for this system and it is actually what made the name of Kanban known (Kanban stands for instruction or visual card in Japanese.)

These cards are typically physical cards that will be placed near or on a product, and then they will travel on through the system in order to provide all of the necessary information to the people who are along the line. There are several options that come with these Kanban cards, and the one that you use is going to depend on the information that is needed at any point in your process.

There are six different types of Kanban cards that you are able to use in your system. These may include:

The conveyance or withdrawal cards

These are the cards that the system should use when it is time to alert people when a part has been completed in one area. In a production line, the parts need to be worked on in one area for a time and when they are done there, the part will be moved over to the next area. This card is going to signal that the parts are ready to be moved to a new area and be worked on there.

When the next station that goes along in the process is finished with the product, they will then take that withdrawal card and send it along to the previous area. This also signals that they are ready to take on additional work because they are done with that first part.

Supplier cards

This is a unique card, one that doesn't always show up on a normal production line. This one allows you to include the suppliers on this system, which can sometimes increase efficiency and ensures that your communication stays open. With this kind of card, you will send your supplier a notification when you will need a new part. This ensures that the message goes automatically to your supplier, without someone having to physically do it, and just makes the process more efficient. While it is possible for the supplier cards to be physical, they will often be digital and can even be represented through a phone call.

Emergency cards

These are the cards that will show up when something has been broken or there is a defect. When the problem is discovered with one part or a series of parts, this card is going to be sent on to the previous station, alerting them of the problem. You can also use these kinds of cards to notify others of the problems that exist so that they make sure the current items that are being worked on don't end up with the same issues. Sometimes these cards will stop the previous workstations from doing work so they don't create any backup of work while the team is working on the issues.

Express cards

These express cards will tell the team that there is a shortage of a part and that part is needed immediately. This will be the card that is used in order to signal that some part of the manufacturing process has slowed down or come to a stop if the parts aren't supplied right away.

Production cards

The production card is going to have a complete list of all the parts that are going to be required at a given time. A workstation can provide this card to any of the areas in the whole facility to tell that area what things they should focus on to continue with production in a timely manner. These cards can also signal the start of production in that facility.

Through cards

This is the type of card that is a combined withdrawal and production card. It is going to be done between two production points that are different, that work together closely. It can save some time because you won't have to send two cards back and forth every time there is a change in the status of a product.

Chapter 5: Using Kanban in a Supply Chain Environment

As we have discussed a bit in this guidebook, Kanban, as well as some of the other parts that come with the Lean methodology, can work to make many processes better. Supply change management, or SCM, is used in many companies, even ones that don't use the ideas of Lean. But as you will find in this chapter, Kanban can actually be used along with SCM in order to make the process more efficient and easier to use. Let's take some time to explore the ideas and principles behind SCM, and how Kanban and other Lean principles can help to make it better and more efficient overall.

What is Supply Chain Management?

The first thing that we need to explore here is what supply chain management is all about. SCM is going to be the management of the flow of services and goods and it will include each and every process that is used to transform raw materials into the final product that goes out to the customers. When it is used the right way, SCM is going to involve the active streamlining of the supply side activities in the business.

The hope here is that this is going to maximize the amount of value that the customer places on the product and it can help the company gain an edge over the competition when they bring the product to the market.

There are a lot of different parts that come with SCM, but it often represents an effort by the suppliers in order to develop and then successfully implement a supply chain that is both economical and efficient at the same time. Supply chains and how they work will vary between one company and another, but they will cover everything from the production all the way to product development and to the information systems that you need in order to help you through this process.

Typically, the goal of the SCM is to centrally control or link together the whole process of a product from production to shipment and then distribution. This is going to allow a company to cut out all of the unnecessary costs while still delivering products out to their customers in no time at all. The way that this is done is by keeping tight control over the different parts that happen during production, such as the internal inventories, the inventories of your vendors, and the sales and distribution.

SCM is going to be based on the idea that pretty much every product that ends up on the market for the customer to purchase results because of the efforts of the different teams or organizations that make up the supply chain. Even though the idea of a supply chain has been around since the first product was created, it has been more recent that a company has paid more attention to them as a value-add to the company.

But what is the supply chain? This chain is basically the network of resources, technologies, activities, organizations, and individuals that are involved in the manufacturing and then the sale of a product. This chain is going to begin when the manufacturer receives the necessary raw materials and then the chain will end when the complete product gets to the customer.

SCM is going to oversee each of the steps that occur from the very beginning to the very end. With so many steps that happen in this supply chain, you can find many places where you can either increase value or decrease value. The right kind of SCM plan in place can make sure that each step is as efficient as possible, which increases revenues, cuts down on some of the costs of the company, and can have a positive effect on the bottom line of a company.

Let's look at an example of how you can add in SCM to a business. Walgreens Boots Alliance Inc. decided to place a focused effort on transforming the way that it manages its supply chain during 2016. The company is known to operate one of the largest chains of pharmacies in the United States, and because of this, it needs to be able to manage and revise efficiently the supply chain to make sure that it can always stay ahead of all the new trends while adding value to the bottom line.

By July 2016, Walgreens had decided to invest more in the technology part of the supply chain. It then chose to implement a forward-looking SCM that was responsible for taking the right data and using analytics in order to forecast the purchasing behavior of customers. This information is then going to back up the supply chain in order to meet the demand that is needed.

An example of how Walgreens would use this is to anticipate the pattern of the flu for that year. When the company is able to anticipate these patterns, it is easier to forecast the needed inventory for any over the counter flu remedies, which can create a supply chain that has little waste. Because of the SCM that is in place, the company is able to reduce the inventory that they have left over and they can reduce the extra costs of having that unneeded inventory, such as transportation and warehousing costs. This saves them a lot of money overall, while also ensuring that they are ready to take on the flu for their customers.

The Principles of SCM

Now that we have gotten a chance to look more at supply chain management and all of the parts that come with it, it is time to take a look at some of the principles that are important with this method. The way that your business will use these principles is going to vary based on what they are hoping to accomplish, the customers they are trying to reach, and the products that they want to sell. The main principles that come with the ideas of SCM include:

1. **Adapt your supply chain to meet the needs of the customer**

Both professionals of the supply chain and business professionals are trained to focus on the needs of their customers. In order to make sure that you get a good understanding of the customer, it is best to divide them into different groups. The most primitive way to segment out your customers is known as the ABC analysis. We will talk about this kind of analysis in the next chapter, but it is basically a method that will group customers together based on their profitability and their sales volume. Segmentation can be done in other ways as well such as by trade channel, industry, and product.

While it is a good idea to meet the current needs of your customers, you should also consider how you can expect the needs ahead of time. You may find that your customers don't really know exactly what they want until the competitor offers it to them. Being ahead of the game and making sure that you anticipate the future needs of your customers can be just as important.

2. Customize your network of logistics

When you are done segmenting your customers based on the needs they have, you may also need to make some changes to the logistics networks to ensure that you are serving each segment. This isn't something that you need to do in each situation, but it can be something that you need to focus on.

3. Align your demand planning so it works evenly throughout the supply chain

Someone who practices the ideas of SCM will learn to share the demand data with all of the suppliers and other organizations that they work with. This ensures that no one in the group has to hold onto any stock that is unnecessary. This principle holds true, but it is found that very few companies are actually following these ideas.

But in one paper that was written by Williams and Waller in 2011, it found a few interesting facts. First, if you make the forecast for demand based on the customer or SKU level, working with your historical data of the order is going to be much more accurate than working with the point of sale data that retailers provide to you. But if you make the demand forecast based on the store level, then the point of sale data from the retailers is going to be more accurate compared to using your own data.

The implication with this one is that the absence of demand sharing isn't always a bad thing. Some companies find that it is beneficial, and others find that they can do just as well, if not better, without it. But if you do get this demand data from some of the companies you work with, make sure that you use it the right way to ensure that you get the best results and the most out of it.

4. **Differentiate the products that are close to your customer**

One example of working with this idea is found in Dell. Dell will keep the components and then only assembles them once the customer puts in an order. This helps to increase how many products they are able to offer, without having to store inventory or worry about the extra costs. This is a strong principle, but you may also want to consider another principle in your business.

Standardization is basically the opposite of differentiation. For example, there are some manufacturers of cosmetics that will formulate the products they sell, and then they will choose labeling and packaging that works with the regulations in various Asian countries. This allows them to make just one product and its packaging that can be sold in 15 different countries.

While differentiation can be nice and can attract more customers in a tight market, standardization can help drive down the costs because of the economy of scale. Depending on the type of business you run, standardization can be a great tool to help you be successful and can reduce waste and costs at the same time.

5. Outsource in a strategic manner

If you are going to outsource your product or any of the work, you need to make sure that you do it in a strategic way. This is a principle that is going to stand the test of time. The best rule of thumb to use here is that you should never consider outsourcing your core competency in the process.

6. **Make sure that you have an IT that will support decision making on multiple levels.**

If you search on Google and try to find the term that says "critical success factor ERP", you are going to find out a ton of information about the way that you can implement this method in a successful way. No matter what, it is best to make sure that an IT project isn't done in isolation. Business process re-engineering is definitely something that should be done before you even get started on implementing one of your IT projects. This will ensure that you have a good understanding of all the deficiencies in the process and then you can use that information to figure out exactly the type of software and technology that is needed.

How Kanban Can Work in Supply Chain Management

As you were reading through a bit of the information above about supply chain management, you may have already seen a few ways that the Kanban system is able to work in order to improve the SCM of your team. Kanban can make sure that you have a visual representation of how this system should go, can give you an idea of the best places to work on improvements, how to avoid issues with the system, and more.

With supply chain management, you are working on ways to improve the system. You will take a look at how the system works from the very beginning to when the product reaches the hands of your customers. But there are various ways that the management of your supply chain can be handled, and not all of them are going to reduce costs and improve the system.

Every company that makes a product or provides a service is going to have some kind of supply chain management. Even if they aren't actively managing that supply chain, it doesn't mean that they are missing out on doing supply chain management, it simply means they are doing it in the wrong manner.

Kanban is a way to ensure that you are actively taking a part in your supply chain management. Instead of just sticking with the current method that you are in and hoping that everything works out well, you can reduce the inefficiencies, increase your profits, reduce costs, and see a great improvement in how well you can manufacture the product and get it into the hands of your customer.

If you have found that a lot of time and money has been wasted when it comes to your supply chain and you want to manage it and get things on track, then there is no better way to do it than implementing Kanban into the system instead. This method, along with the techniques that we have discussed in this guidebook, are going to make a big difference on how much success you can see and the results that you get when you decide to get started with SCM.

Chapter 6: What is the ABC Classification?

On this guidebook, we have spent some time talking about inventory and how it works in the Kanban system. With many traditional forms of business, having a lot of inventory stored up to meet demand was a part of the game. This is how many businesses would make sure that they had enough products for their customers. It could make the process faster, but often it ended up adding more risk to the process because they could easily be off with their predictions and then they end up with a ton of inventory that they paid for, and they pay for storing, that the customer doesn't want.

With Kanban, inventory is not used as much. The idea is that the inventory is not an asset, it is a risk and a waste because of the costs that come with it. This method can sometimes run into trouble if the lead times are too long, but that is something that has to be worked out when it comes to using this method. With the right suppliers, the right plan, and everyone using the visualization, it won't take long to get the product to the customer on time without all the wastes.

One method that can be used to help manage the way that the business is run, to determine if there needs to be any kind of inventory and more is the ABC classification. This kind of classification is going to be a type of ranking system that can identify and group items based on how useful they can be for achieving your goals in the business.

This kind of system is going to take the different aspects of the business and group them into three categories. These three main categories are going to include:

A: Extremely Important
B: Moderately Important
C: Relatively Important

The ideas that come with ABC classification are going to be associated pretty closely with the 80/20 rule. The 80/20 rule is going to be a metric used in business that says 80 percent of the outcomes for the business are determined by 20 percent of the inputs.

The goal of using the ABC classification is to provide ways for a business to figure out what makes up that valuable 20 percent so that they can then control the segment as closely as possible and maximize it as much as possible. Once all of the segments have been identified by the business, each one is going to be handled in a different way. The most attention is going to be given to A, less attention will be handed to B, and then the least amount of attention is given to C.

ABC classification is often something that is going to be associated with controlling inventory, but it is also used to help rank the different customer segments, but segments of the business can cause the most risk financially, which employees are seen as the most valuable, and where it is likely that a bottleneck will become a problem.

Defining the ABC's

The first thing that you need to do in order to get started with an ABC analysis is to separate out your products or customers into each of the different categories. Remember that we have A, which includes the items that are the most valuable to the business and therefore need the most attention, then we have Class B and Class C. Let's take a look at how we should separate out each of these in the ABC analysis.

Class A Inventory

The items that fit into the class A are going to be about twenty percent of the inventory at most. But they are so important because they account for at least 80 percent of the profits that the business is able to make. Some of the characteristics of this classification include:

1. Items in this section are going to cost the most to use. This could be because the unit is higher in cost or the volume usage is high on an item that has a low unit cost.

2. These items can present a bigger risk. Because these are necessary items with a higher expense, these items need to have a lot of human capital invested in them. Most inventory managers need to spend the majority of their time and attention on replenishing these items. Without the right stock levels of these, there can be a lot of consequences.

3. These items are going to be replenished using a sawtooth curve. Because the costs that come with these items are higher, there are fewer on hand, even though the orders are placed on a regular basis. This means that a low lead time and standard costs are important on these

4. The lead time on these should be somewhere between three to ten days. These are going to comprise the majority of the on-hand value for a company.

Class B Inventory

After you have figured out the most important items to put into Class A, it is time to work with the Class B items. These are going to account for somewhere between 15 to 30 percent of the items for the business. These are going to have a lower impact on inventory spending, around 15 percent for the year, but are still things that need a little bit of attention as well. Some of the characteristics of Class B include:

1. These types of items are going to require some standard costs that need to be paid at predictable intervals. These items are going to be needed for production on a regular basis and they will be on a routine for ordering, but won't be needed as much as class A items.

2. These items are going to be important enough that the team should manage them, but they don't need the high level of control that the class A items do.

3. These items should be on hand, but keeping them at moderate quantities should be fine. Keeping as low of an inventory as possible is the goal and you should limit the number of transactions wherever you can.

4. The typical target lead time for these items will be between 5 to 15 days. The items that fit into this category will contribute somewhere between 15 to 25 percent of the inventory value.

Class C Inventory

Finally, we are going to move on to the items that fit into class C. These are the majority of the inventory stock, with at least 50 percent of the items fitting into this category. Even though there is so much in here, these items are only going to account for five percent of the amount that the company is going to spend. Some of the characteristics of products that would fit into this category include:

1. These items aren't going to cost a lot, but their consumption is very high. This could be things like stampings, literature, springs, and such. However, there are times when a class C item is going to be more expensive, such as a sub-assembly that may be $700, but the team only uses a few times a year.

2. The replenishment of these items should be as automated as possible. These shouldn't require any attention from the team because they are then neglecting the more critical items in the other

categories. Because there are so many items in this category, going through and actually working with them, rather than automating it can waste a lot of time.

3. Procurement managers need to focus on reducing how many transactions are needed to acquire these types of items. The tedious task that comes with this kind of work, such as making the order and inspecting it before moving to the right department, can cost a lot more than the direct acquisition of those items.
4. While these items are only going to take up 5 percent of the spending budget, they are going to represent 20 percent of the on-hand inventory dollars because it is easy for companies to overstock these items to avoid shutdowns that come when they run out.

ABC classification can be useful when it comes to handling your business. It ensures that you are going to be able to provide the customer with the products they are looking for, without a huge lead time in the process. You can decide which products are the most important, based on how much they make you, and then work your way on down the line until you reach the products that don't make up a lot of your profits but are still important.

The Steps to ABC classification

While the concept that comes with ABC classification is pretty simple, going through an inventory that is large or complex can seem really overwhelming and tedious. The payoff of this can be really great, even though doing the classification the first time is going to take some work. After you have committed yourself and the team to do an ABC analysis of the inventory, the following will be the steps that you must use to get it done.

Separate the items that are purchased from the ones that are manufactured

There should be two item lists here. One of them will be for the items that the company purchases and the other are for the items that the company manufactures. This helps you to get a good look at the highest, middle, and low costs in each category. Because the annual cost of all those manufactured items often ends up being higher than the purchased parts, often these are the items that are going to take over the limited class A spots. These can push out some of the purchased items that should be done in Class A into the lower categories.

One thing to consider is to count maintenance, repair, and operating items on your list. The expense of these supplies can be easy to forget, but they do add up and you do want to have a good look at the value of your inventory. Remember, anything that contributes in some way to the value stream, it then deserves a bit of your attention.

Collect the annual usage data and the standard cost data for all part numbers you purchased

Now you will need to go through and create a spreadsheet that has three columns. The first column needs to have the part number of each item in your inventory. The second column should have the unit cost for every item. And then the third column is where you need to write out the annual demand that the customer has for each item.

Calculate out the annual amount you spend on all the parts

With that spreadsheet that we talked about before, you need to create another column to calculate the annual amount that you spend. To figure out this number, you just need to take the standard unit cost for the parts and multiply it by the annual demand to get this number.

Now, take advantage of the sort function of the spreadsheet so that you can sort out the rows based on the annual spend. Do this in the descending order so that those who have the highest spend each year have their values at the top. When you have it all sorted out this way, you can use a non-sorted cell in order to figure out the total amount that you spend for all of the parts.

Add in another column that shows the cumulative running total

Now we want to calculate out the cumulative annual spend. This is simply going to be a running total of the annual spend on each item that will accumulate to equal the total annual spend of all the items. For each of the rows, you want the cumulative running total to be equal to the sum of the annual spend of itself, plus the annual spend of all items before it when you sort this out by descending order.

Identifying items that go to Class A

The first thing that we need to do here is to figure out which items should be considered Class A items. To find that, you first need to find the point that will act as a threshold for 80 percent of the total annual spend. To do this, use the following formula:

Target total annual spend threshold = total annual spend for all purchased items X .80

You can then round this to the nearest ones, tens, or hundreds place (depending on how large the number is to start with), in order to get a good estimate of your threshold for the items that fit in there. Now, take a look back at the cumulative annual spends that we worked on earlier and compare them to this target. Find the first row that is greater than this target and let this be the first estimate of your cut-off for determining if an item should be Class A or not.

Now you can move on to comparing the annual spend of the items on the threshold between Class B and Class A. If you see that there are two items that are pretty similar on either side of the threshold, this means that you need to either raise or lower that number. You want to make sure there is some type of meaningful gap between where an A item ends and where the B items start. You don't want the threshold to deviate too much, but a little bit is not a bad thing.

Once this point is established, you can add in another column to that spreadsheet that will house the ABC classifications for each part as you go through. In this column, you can enter A for each item that, based on the analysis that you just did, qualifies as an A item.

Identifying items that go to Class B

The next category we are going to work on is the items that fit into Class B. These are the items that will take up about 15 percent of the amount that you spend each year. To find the threshold that will separate out the items in Class B from the items in Class C, you need to figure out the value of 95 percent of the annual spend. This 95 percent is used because it shows you the amount that both Class B and Class A contribute to the total. But since we already went through and know what the A amount is, we can use this number to help us find the best cutoff between the C and B inventories.

This estimated cutoff can sometimes include the items that you designated to Class A, depending on if and how you went through and adjusted the threshold manually in the previous step. Make some adjustments manually so that the cutoff point is meaningful, just like we did with the Class A items. Then you can designate the right non-A items below the identified threshold as B items.

Identifying items that go to Class C

All the remaining items, the ones that didn't fit in with the requirements that we put in for Class B or Class A can go into Class C. There are going to be way more items that fit into

Class C than the other two combined, and Class A should be the smallest.

Repeat this with all the manufactured items

Now that we know the steps that are needed for doing an ABC analysis, you can repeat the steps for all of the manufactured items in the company as well. Remember that it isn't that unusual for the manufactured items to have a high total annual spend as well. This methodology is still one that is sound and can work, regardless of the amount the annual spend comes out to. It is also the main reason why we went through and separated out the purchased parts from the ones that were manufactured in step 1.

The extremely high cost that comes from manufactured items is often the result of a duplicated demand. When one part is used in a way of fabricating another, then those parts will then be used in an assembly before it becomes part of the finished product. This means that the parts that become a part of the finished product can be counted a few times in the demand column, so sometimes it is hard to figure all of this out.

Chapter 7: Traditional MRP System vs. Kanban

Before companies started to work with the Kanban system, many of them relied on a method of inventory replenishment that was known as Material Requirements Planning or MRP. MRP is going to work by tracking the data in the system and then comparing the inventory that the company has there already, or that they will receive soon, with the inventory that the customer will demand over a certain period of time.

MRP will be able to analyze the supply data and the demand over a planning horizon or a specific time period. The company will be able to choose how long the time period can be, and often it was based on the idea of the lead time and how long it takes to get the raw materials, how long it takes to create the final product, and the current demand of the customer. This data and the demand will then be used in order to decide how to schedule the replenishment of the product.

The orders for replenishment for MRP will be triggered if the future demand for parts will be greater than what this system report is on hand. So, if the company has enough items in the warehouse or on hand to make 10,000 of the product, but the projections for demand are at 15,000, then the MRP order for replenishment would be triggered. The system will then be able to evaluate the data. Depending on the way that the system is set up, it will either lead to an automatic reorders or it will generate an order form that will need approval in order to be used.

The MRP formula was pretty simple to work with, it was simply the following:

MRP order quantity = on-hand balance + Upcoming receipts – future demand

If you end up doing this and the calculation gives you a number that is negative, this means that a new order is needed. The system can either go through and do the work automatically for the company, or it will send out a recommendation that needs some approval first.

The MRP idea is simple to work with, but there was the need for four pieces of information to be collected by the company on a routine basis, and these numbers need to be as accurate as possible. Without these, you will find that the MRP system isn't going to work that well. The four pieces of data that are needed in the MRP system include:

1. Supplier lead time
2. Future demand
3. Future receipts
4. On-hand balance

Let's break these down a little bit and get a better understanding of how each of them can work in your system.

On-hand Balance

This balance is going to be the recorded in-stock quantity of an item. This will be recorded using a specific unit of measure for every item. The quantity is only going to be accurate if the balances are balanced perfectly at each transaction for the physical inventory. Inventory errors can often come in and are going to involve these types of balances because they can be difficult to ensure that every use and exchange of the item is accounted for with the matching update in the MRP system.

There have been some efforts done in order to ensure that all of the items that are damaged, the ones that are used as a sales tool, or the samples from engineering, are also accounted for in this process. Facilities that are able to batch the production process will find that it is hard for them to account for any WIP or the partially consumed inventory.

Another issue that comes with this and can negatively impact this kind of balance is scrap. Work cells will batch the reporting of scrap. Rather than going through and reporting this after each shift, scrap can sometimes be filled in just once a week or only when the bin for scrap has been filled. Unless the MRP system is able to assume the scrap rate accurately, the reported amount that is on hand can be way off from reality.

Future Receipts

Another thing that we need to take a look at is future receipts. These are going to be tied to the purchase orders as well as the manufacturing orders. When there are orders that have been maintained accurately in the MRP, starting with the entry of the order all the way through it being completed, future receipts that are reported in this system are going to be accurate.

However, the issue comes when the MRP reports the wrong due dates or the wrong quantity of receipts because of discrepancies between the order quantity and the actual amount that is on the receipt. There can also be issues because there are differences between the original due date for that receipt and the actual date of the receipt. Even teams that are disciplined and accurate can sometimes fail when it comes to reconciling the MRP until it is all too late.

Future demand

Future demand can also be used by the MRP system for replenishment. Future demand is going to be based on the customer orders that are already known, and it is the best guess of what the demand is going to be. Since it is just an estimation of the material that might be needed soon, there could be a lot of different factors that will go in and undermine it.

Often, you will find that the planning horizon of the MRP isn't going to match up with the amount of time that your chosen demand accounts for. For example, if your production planning process is going to need 8 weeks of demand but the company is only taking care of four weeks of orders, there is going to be a gap in there that the company needs to find a way to account for.

You will also find that the numbers for the MRP can be invalidated if there are some last minute changes to the forecast. This is something that can be true even with forecasts that seem to be really accurate. For example, it is possible that someone on the team is going to share an accurate 12-month forecast, but there could be some volatility that comes up in a shorter time horizon, which can make it more difficult to plan things out.

Because the demand of the customer is going to be entered in terms of the finished goods, orders are sometimes going to cascade down and that can affect a bunch of different areas. Because of this, it is possible in the MRP system that a negligible inaccuracy in the BOM can shut down the whole line if it isn't taken care of properly.

Actual lead time from the supplier

You need to have a good idea of how the lead time of your supplier stacks up. They may meet with the industry standards, they may be a bit slower than industry standards, and they may even be faster. If you do your order due dates based on what is considered the standard lead time in the industry and not the actual ones from the supplier, it is possible that the MRP system is going to generate an order for the right amounts of parts, but it will do so at the wrong times. This can lead to too much or too little inventory, which isn't good for anyone in this process.

MRP is the Push

The origins of the MRP that we have been talking about in this chapter date back to the 1960s, just a bit after computerization of the nuts and bolts of the logistics of manufacturing. It worked well enough for a number of years and helped companies to get the product out the door and into the hands of their customers. But it wasn't always the most efficient, and sometimes, it cost more than companies were able to keep up with.

The ideas behind MRP will be push mechanics. With this push mechanics, there are going to be a few principles that need to be followed including:

- Demand forecasting: The push systems, like those found with MRP, are going to guide production and are based on predictions into the future about how much demand there will be for a finished product. The math to come up with these forecasts will take both the existing inventory of the products that the company has on hand, as well as the raw materials that are consumable into account to get a good number.

- Continuous production: Because the production in the MRP system is going to be guided by the forecasted quotas, rather than the on the fly orders that come down from the supply chain, the finished product is then going to be added to the inventory, no matter what the demand is at that time.

- Bill of materials or BOM: We discussed this one a bit earlier on, but it is going to include a list of all the materials that need to go into a finished product. This can include all the parts, the components, and even the raw consumables for the product. This list can be helpful in case something breaks, something new needs to be ordered, or if there is any other reason that someone needs to see all of the various components of that product.

Kanban is the Pull

On the other side of things is the system that we have been discussing throughout this guidebook. Kanban is more of a pull system that was started in Japan when Toyota was starting to look for some ways to increase the amount of efficiency in its factories. One of the employees there, Taichi Ohno, was able to draft the system that was used by Toyota based off some of the observations they had on the way that supermarkets were managing their inventory.

In the supermarket, the shelves were only stocked with the amount of goods that a store was able to sell. And the customers wouldn't take any more items than the amount that they needed and no more. Kanban and the other options of Lean will follow this basic idea, and this is why they are known as the pull manufacturing techniques. These are going to be based on the following principles:

- Kanban tickets or cards: In the Kanban system, the actions are going to be driven with the help of the Kanban cards. These are signals that say that a supply of a specific resource should be replenished. So, if a factory needs to come up with 25 widgets but they lack one of the parts that are needed, the assembly line would take the needed Kanban card, move it to the supply desk, and communicate that this part needs to be restocked. This means that nothing along the line is made or created unless there is actually a demand for it.
- JIT production: JIT is the idea of inventory and supply underpinning this kind of manufacturing. It helps to tighten up the efficiency of production and can make things continuous to improve the system. It believes that any surplus of product and parts are a type of inefficiency that should be avoided in order to keep the costs down.

Which of These Options is the Best?

Despite the idea that both of these methods have been around for over 40 years at this point, the debate of which one is the best system is still up in the air. The trend right now is to go against the ideas of MRP in favor of Kanban because many companies are looking towards the Lean methodology to improve their functions and to ensure that they are able to reduce the amount of waste that is produced throughout the company.

With that said, there are some flaws in the Kanban system. This can be really apparent when it comes to seeing any big disruptions in the demand or the supply. Often, these disruptions can be handled if there was a little bit of inventory surplus present, but the Kanban system doesn't allow for this at all. But in return, this system ensures that the company doesn't keep around any more products than it needs, which can help it be more efficient and reduces the costs that come with making and storing the extra products.

Kanban is often seen as the most efficient out of these two methods, even though there may be a few pitfalls that come with the system. It is easy to use, it saves on costs because the company doesn't have to make and store items that aren't in demand, and it helps the team come up with ways to deal with the various inefficiencies that are present throughout the system. The process is pretty simple to use and there are various diagrams that can be implemented to ensure that it works properly.

With the MRP method, the company is going to try and predict the amount of demand there is for the product and then pull it in so that the product can be made to meet the demand. This can be efficient and can get the product out to the customer faster than other methods because often the product is already created and ready to go.

The speed times may be faster, but there are a lot of inefficiencies here. The company may end up predicting wrong and then they have a lot of inventory that is left unused and that will hopefully be sold in the next time frame. This means that in that time frame, the company took on more costs to make a product that didn't even sell. And they have to store that product for a number of weeks, which can be expensive as well. Because of these issues, the traditional MRP system is often seen as one that is inefficient to use.

Kanban, on the other hand, is a JIT methodology. When a customer orders the product, it is then manufactured and sent out. The product is not made ahead of time, and there isn't any worry about storing it properly and ensuring that you get it sold. This saves a lot of money for the business. And if it is done right and with the right supplier, lead times will be kept to a minimum so the customer won't have to wait a long time to get this product to them. This method is often seen as the one that is more efficient and more cost-effective, and that is why a lot of businesses choose to implement it in our modern business world, rather than the traditional MRP system.

As you can see, there are some benefits and some negatives to using each of these types of systems. It is important to understand how each works and why you may want to work with one over the other. Many companies are starting to drop the MRP system because they find that it is inefficient or it just doesn't work for them, especially if they are trying to implement the Lean methodology into their production process.

Chapter 8: Designing a Kanban System

When you create a product for your customers to purchase, you will notice that the balance between supply and demand is always going to be changing. There will be times when the demand is more than your supply and other times when the demand will go down and be lower than your supply. However, in many industries, it seems that the supply of a product is steadily growing faster than the demand for that same product.

As more and more products get into the market and it becomes more saturated all the time, you will find that customers are becoming harder to convince when it comes to selling them your product. Many of them are looking to get a lot of value out of the products they purchase, mainly because they have a lot of choices out there from your competition.

The good news is that Lean is a great option to use in order to get yourself ahead of the competition. There are a lot of options that you can use in order to get to this, but value stream mapping is a method that can work with the Kanban method we've been discussing in order to help you get a visual representation of what should happen in your product cycle.

Let's take a closer look at Value Stream Mapping or VSM, how it works with Kanban, and the steps you need to take to make this work for you.

What is Value Stream Mapping?

VSM is a process that will make it easy for you to get a detailed visualization of each and every step that occurs in the work process. It is going to be a good representation of the flow of goods, starting with the raw materials that your supplier brings you, all the way until the finished product reaches the hands of your customers. For example, the value that a software company is able to deliver to the customers will be all of the software solutions, as well as the features that are found inside.

This value stream map is going to display all of the steps of this work process. It needs to include each and every step of the process, from start to finish, and it should all be written or drawn out. This allows you a way to visualize each task that the team works on and can provide a single glance status report about where each assignment is in the progression.

One note here is that in the Lean methodology, value is going to be everything that your customer is willing to pay for.

However, when you are working on the VSM, there may be some steps that you include that don't directly bring value to the customer, but these steps still need to be added in because, without them, the final product would never be completed. A good example of these kinds of steps would be the quality inspections. These inspections are important to ensure that the product is done on time and that it meets the high-quality standards of your customers and of the market. But these don't really add value to the customer and they aren't going to pay for these inspections. However, if you send out a product that doesn't meet the quality standards of the customer, the customer won't be happy and they are less likely to purchase from you in the future.

The main purpose of going through and creating this VSM is to show you and your team the best places where you are able to make improvements in your production process. When you have to write out each step and look straight at it, it is easier to figure out what you can do to get rid of wasteful steps and how to add more value into the product.

To make this work, you simply need to put on display every step that is important to the workflow. And then you need to evaluate it and determine how it is going to bring value to the customer, either directly or indirectly. This will make it easier for you to do a deep analysis of the process and can make it easier for you to see the glaring wastes and extra steps that you can get rid of.

So, where did the idea of VSM come from and why is it so popular in many businesses today. The idea of value stream mapping became popular in the second part of the 20th century with the rise of the Lean methodology and all that it entails. Lean was the biggest foundation that was made with the Toyota Production System. At the time that this came out though, the idea of VSM was a part of that production as well.

Toyota was not the one who invented this practice at all. In fact, the idea of mapping out the workflow in a more visual way was something that was shown in diagrams in a book called "Installing Efficiency Methods" that was released in 1918. By the 1990s, this process became a big part of the production process in big corporations. The popularity of this continued to grow and soon it started to spread out to the manufacturing world and into other industries such as marketing, IT operations, and software development.

One area that seems to really be growing when it comes to value stream mapping is that for knowledge work. This is because the VSM can allow teams to work together with a visualization of their work, which allows them to collaborate better than before. Even each individual contributor is able to take a look at the VSM and get an idea of how the team is progressing in the work that they need to get done.

Because of this, the team is able to increase how efficient they are and how well they can do work handoffs. Work handoffs are one of the biggest culprits of accumulating wait time inside the system. And since waiting is considered one of the major wastes that are found in this system, it is important to limit it as much as possible.

When you use the idea of VSM, you are able to map out your process from the very beginning to the very end. This kind of chart is able to help you get a visualization of where the handoffs are actually occurring in the process so that you can discover where any type of bottleneck is going to occur in the process. From there, you can come up with some methods of reducing the damage that this can do to the productivity of your team.

How to Create a Value Stream Map

At this point, we have taken a look at what a value stream map is about, but now it is time to divide it up and actually create the map to suit your needs. It isn't going to do you much good if you have no idea how to create the value stream map. How are you going to get it set up and ready to use and get the most use out of it?

The first step that you can do with this process is to calculate out the takt time. If you don't remember how to calculate the takt time, go back to our previous chapters to review how to get this done. In the example we are going to use, we will have a daily demand for production for 700 pieces. The other information that you will need to make this happen and to figure out the takt time include the following:

Days per week: 5
Shifts in the day: 1
Break minutes for each shift: 30
Hours per shift: 8

Take the time to punch out these numbers and figure out what the takt time is. We are going to skip ahead here and say that takt time will be 39 seconds for each piece. This means that you will need to produce the product every 39 seconds in order to get the customer demand handled.

The next thing that you can do is get out a pencil and maybe a few erasers so you can make sure the steps are all in place. The best value stream maps have to be redone many times before they look perfect. Never start out with a pen because you are going to miss things, need to move things around, and more. This is just asking for a bigger mess than it needs to be.

From here, you can bring out a piece of paper to write on as well. Go for a size that is bigger than printer paper in order to get things done. 11 by 17 is a good size because it is big enough to do the work, but still small enough that you can carry around. With this method, remember that we are working with the pen and paper version. Even if you plan to use software with the VSM, it is often a good idea to draw out the map by hand to get a good feel for it.

Now that you have your supplies, it is time to walk the process front to back. You and your team should walk through the whole process to get a good feel for the general flow. This is also the time for you to define the starting and the stopping points of the process. Make sure that everyone is taking on too much work at any time or this will create the bottleneck that we are trying to avoid.

Here we can draw in the details or the customer box. This is going to be on the top right hand of the paper. Make sure that you draw out a small saw topped box that is going to be the representation of the customer in this scenario. In addition, note out the daily or monthly demand and the takt time that we calculated earlier in this section.

You will then want to start at the end of the process. Draw out your map going from the back to the front. And if you need to use that eraser, that is fine as well. Pick one person on the team who is going to be your scribe and who will draw out the map for the whole team.

If you would like to make sure there are no mistakes in the process, you could have everyone help out. Ask each person on that team to make their own little map. Then you can compare the different parts and consolidate them when everyone gets back together. Or you can divide and conquer, asking everyone to work in a smaller group and just take on one part of the production process. Some may work on the start, some on the middle, and some on the end of the process.

As you can see, there are different methods that you can use when it comes to mapping out your visual stream. You need to experiment a bit in order to find the method that seems to work the best for the team.

Next, you need to focus on the material flow first. Focus on the bottom portion of the map, including the data boxes and the process boxes. When it comes to the data boxes, you don't have to stop and put all of the data in there perfectly that day. Just try to fill it out in the best manner that you can. You can always make it a goal to do some homework and validate the figures or add more in at a later time to finish up. Even if you think you have data that is solid, go through and validate these measurements to ensure that you are putting the right ones into the program.

From here, you need to add in the inventory and the wait times. Once you have these data boxes, it is time to add in your waiting times or the inventory. For inventory, you just need to count out the number of pieces that show up in between the processes and then note them in the system. You can also spend some time here converting these pieces into days' supply. To do this, we need to divide the number of pieces by the average daily demand, which is what we used in order to come up with the takt time.

So, if you have a daily demand for the product of 10 pieces and you find that you have 20 inventory pieces that show up between step A and step B, this means that you have enough to do two days of work between these processes. We need to note this on the timeline, but we will add it in later.

Of course, you shouldn't waste your time mapping out each and every part number. Pick out one or two components that are key and then start with those. You can always add some more to the map as you need later.

Now it is time to move on to drawing out the information flow. This is the main step that is going to separate out the VSM from other process maps. You will be able to use this in order to understand how the material is going to flow through the business and how to get production done.

During this step, we are also going to draw in the control box for production. For many, this may include MRP because most mass production systems are going to see several straight lines, which means manual information that comes right out of the MRP box that is aimed right at the process step box.

And to finish up, we need to add in the timeline. This should be near the bottom of the map. This is a line that should be a sawtooth that will help you to separate the value of added cycle time, which you can get from your data boxes and from any time that doesn't add value, such as the hours' or days' supply information. Then you can sum up all of the value add cycle and note them near the end of their timeline. You can also note your inventory time as well.

How to Map a Value Stream Using Kanban

Now that we know a little bit more about value stream mapping, it is time to explore how you can do this with the use of Kanban. Kanban is one of the best and more reliable tools out there when it comes to doing your own value stream map. This is especially true when you want to be able to visualize a knowledge work process. You will find that Kanban with the VSM is really easy to understand and there is the added benefit of being able to make your changes easily.

In order to map out the process for making your product from beginning to end with the help of Kanban, you will just need two elements. These elements include a Kanban board and then you need some cards to hold the place of all the assignments the team is currently working on.

For those who are newer to the idea of doing visualization of the workflow, it is much better to start out with the VSM on a small basis. Start by doing this with one team or one department to get the hang of the idea, rather than trying to make it work with the whole company. In addition, when mapping the workflow, make sure that you choose the method that is the most convenient. You can choose from either a Kanban software solution or a Kanban board.

After you have made that decision, you can then gather up your team. This meeting is a time for you to explain to everyone the purpose of doing the VSM and what you want to achieve when you start to implement it in your business as well.

Once you are sure that everyone has had their questions answered and they are all on the same page, it is time to bind the process together. Because we are working with Kanban, the board will automatically come with three work item states that you can choose from. These work item states include To Do, In Progress, and Done.

The first focus that you should work on is taking the In Progress category and breaking it down into a lot of different columns, including ones that will represent the steps that are the most critical when it comes to your workflow. For example, if you run a software development company, some of the steps that may be present in your business and which should be shown on this VSM include code review, testing, coding, and tech design.

Of course, this is just a brief look at what you may find there. You can add in as many different columns as you would like and as are needed for the Kanban board to be complete. More precise mapping is of course going to give you the most comprehensive view of the program and then more indications of where you are able to improve your performance overall.

You can then pay some special attention to any of the queue stages of the process where there are going to be handovers of the work occurring. This will help you see exactly when the work is going to change hands and you can then focus on those later on to determine what changes you can make to reduce the waste or the errors that happen there.

When you've got a bound process, make sure that you select a range of performance metrics that you would like to monitor. Some of the performance indicators that you will want to pay attention to in this place include:

- The amount of work that is currently in progress
- The throughput of the system
- The lead time versus the cycle time for any assignments that are being done.

As we have discussed a bit in previous chapters, you should always work to make the cycle time and the lead time as close to each other as you can. Even if you are able to keep efficiency present when processing work, having a backlog that is too long means that your customers will have to wait a long time in order to get their order. And this is something that no business wants to deal with at all.

The next thing to look at is the throughput. You must make sure that the throughput for the process is as high as possible.

But then you need to also worry about the quality of the product and the way that it delivers value to the customer. If you make the throughput too high, then the quality may go down. Try to strike a good balance between the two of these to get the best results overall.

And finally, make sure to watch the work in progress. This work in progress factor is going to be a big indicator when you are using Kanban and Lean because the more work that is in progress at a time, the more wait time that will start to accumulate. Be sure to regulate the assignments that the team is going to take on and how much they do at one time, and then make some adjustments based on the data you have from the workflow.

Doing this process is not something that you focus on once and then avoid it or assume that the workflow is going to stay the same. Your business is going to progress and change over time, and it is important for you to review and inspect these KPI's as often as possible. This ensures that you can speed up the delivery time without forsaking the quality of work you provide to the customers.

There are a lot of different things that you are able to do when it comes to working with the VSM. When you have everything lined up and ready to read off a chart or a diagram, it is easier to see exactly what happens with each step of the process. Sometimes, just being able to see the way that the process goes and looking at all of the different steps right out in front of you, can make a big difference in how much waste you are able to get rid of.

When you combine Kanban together with the ideas of doing VSM, you can get even better results. You will go into making this kind of map with the idea that you want to get rid of the waste that is present in the production cycle. You will look at each step, figure out what is valuable and what isn't, and then make the right changes to make the system more efficient. No matter what kind of business you are running, you will be able to see some benefits when you create and follow the VSM.

Chapter 9: Maintaining Kanban

Kanban Card Maintenance

The cards in Kanban are meant to cut out the waste, reduce the time that people are waiting, and to make sure there is an improvement in collaboration on a project. But these cards aren't magical. They won't be able to do everything on their own. You need to be able to maintain them effectively in order to get the full potential out of them. But the next question here is how you are able to maintain the Kanban system. The five ways that you can do this include:

Immediately remove any tasks that have been completed

When the operations of a business are in progress, this kind of system can be very useful for helping anyone on the team track the status of the work tasks, delegating jobs, and then collaborating as a team in order to produce high-quality results. But when you use these boards on a regular basis, it is important that you always work to keep them up to date.

If you keep completed tasks on the board after they are done, it makes the visualization of the Kanban board harder to work with. Your colleagues, as well as yourself, will then have to spend more time looking through the items to figure out what still needed to be done. This slows everyone down, rather than making their job easier and convenient. When a task is done, it should be removed out of the system right away so that the whole board is always up to date.

Color coding

The main purpose of using the Kanban system is to allow you some easy access to information about the tasks that are in progress or need to be done. You will find that working with some system of color coding can make it easier to look through the Kanban board effortlessly. Whether you use a physical or digital system, contrasting colors can make sure that urgent jobs will stand out in the system. You can also use different colors in order to keep track of the progress of the team members' and ensure that your projects are completed punctually.

You can pick out the colors that you would like to use when it comes to setting up the board. For example, you could have red be for "to do", yellow for the projects you are doing, and green for what is done. This makes things simple and straightforward to make the board work well.

Move the cards between the columns as needed

If you want to make sure that the Kanban board is used the proper way, you need to make sure that the cards are moved through the system when they reach a new level. It is likely that you will move around the cards more than you will create the new ones. If you use an app, it will only take a few seconds to move the cards around, moving them off the To Do list and over to the In Progress list, and then from the In Progress list to the Complete list.

It only takes a bit longer in order to move the cards from the physical board though. As soon as the status of a task is going to change, just get up and make a change. Moving cards around on the physical board is not going to be as effortless as using the app. But if you use magnets and sticky notes, it only takes a few seconds as well.

Add in notes when needed

When you use this kind of system as a way to track more than one project, not putting in the details that are needed about a task could add a lot of confusion and can waste time. In addition, vague information is just as bad. When the employees start to become muddled when they complete tasks, they end up making a lot of mistakes.

One thing that you must always do when you have this kind of system, especially in the digital system, is to add notes to the cards. A physical board just needs to have cards that are big enough so you can add in some details. List out as many details as possible with the cards so everyone knows what needs to be done with that system.

Attach the necessary checklists

Both the physical and the digital Kanban boards will support you using a checklist in order to track the progress of that operation. This checklist can be nice because it is a simple and efficient way to ensure that everything is getting done, which means that it is a good idea to implement this in your own Kanban system.

If you are working with the app, you will be able to access this feature in just a few seconds. You can then go in and create your own checklist for any tasks that you need. This is nice because it allows you to break one big project down into smaller tasks, while only using up the space for one card, rather than many on the board.

With the checklist, make sure that each project is checked off as soon as it is completed. This helps you to avoid risking any wasted time working on a task that was completed earlier by someone else. The checklist is a nice way to keep everyone on the same page and to ensure that you are going to get all the work done without a lot of wasted time.

Kanban Audits

Once you have implemented and started the Kanban system, the next task is to make sure that this system keeps going and that you are able to reduce the Kanban quantities. The task of keeping this all going will center on the process of auditing and then making any corrections that are needed as the problems are discovered.

When we talk about auditing the Kanban system, we mean that we are doing some tasks that ensure your Kanban system is running the way that it was designed to do. The schedulers or the material managers for the plant will usually be the ones who will go through and perform these tasks. The auditing process is going to consist of both cycle counting and then a review of the production records of the past. The cycle count is going to provide you with information about the inventory that is up to the minute and which you can then compare to the signals to ensure that they do match.

This review over the records for production will be needed in order to confirm that your production operators are actually following the signals that Kanban sends out. Any time that the audit finds there are some problems, the right changes must be implemented or they need to go through and get help to fix these problems. If the correction to that problem is not obvious, problem solving is now a requirement in order to find and correct that problem.

This auditing function is also going to include ensuring that the Kanban size is adequate in order to support the needed changes to production when they come up. To determine whether this is true or not, you should compare the current forecast for the production requirements to the base-line quantities that are used to size the Kanban. If these requirements end up changing by 15 to 20 percent, then it may be time to consider resizing the Kanban.

The auditing process is not one that a lot of people enjoy and it sometimes seems like a very mundane process, but often the reason that a Kanban fails is because no one was auditing it and a minor problem or a few problems ended up sliding over time and making things worse. As a result, the problems of the Kanban never got fixed, and the production requirement changes were never addressed in a timely manner.

When all of this happens, the Kanban that you worked so hard to implement will become useless and the company will stop using them. Because of this, the Kanban is then left aside and the company will go back to the push or the original forecast schedules from before.

Formalizing the auditing process

The first step to this auditing process is to set up a schedule that you can maintain when it comes to doing the audits. You should make it a part of your daily routine in the beginning to get used to the process. You can later reduce this to just a few times a week. Remember that the main purpose of doing the audit is to help get this process up and running and to ensure that it keeps on running.

When you work on auditing your Kanban, you need to look and see if any of the following items are present:

- Do you see that any of the scheduling pieces or signals disappeared?
- Is the inventory correct?
- Does anyone in the process seem to have any concerns or questions?
- Are the customers, material handlers, and operators sticking with the design?
- Do you see that the original assumptions on sizing and your original calculations still apply?

To determine if one or more of these situations exist, the auditor must first look at their Kanban. You can't sit in the office and get this done. You have to actually get out into the production line and see how things are working. The audit is going to take the physical form of counting the signals, doing a cycle count, talking to the customers, material handlers, operators, and supervisors about the system, and checking the Kanban to see if it is still meeting with some of the long-term requirements that are set.

Finding where the problems are

The next question is how will you be able to detect these kinds of changes? The audits will be able to help you uncover these issues. In terms of the audit, you will start to see that the material stops flowing, the production flow will stop following whichever preferred sequence you have and will jump around a lot, and stock-outs will start to happen on a regular basis.

If you start to see that there is some kind of issue in the process, then it is time to go through and investigate what is causing this. Don't consider failures in the process discipline as the same thing as a failure in the scheduling of Kanban. Don't stop scheduling on Kanban just because you find some kind of problem in the system. Actually take the time to make sure you find the cause of the problem and fix that instead.

One technique that you can use when you find that there is a problem is to ask yourself and anyone else involved, what has changed? The '5 Why's and 1 How' approach is a good way to determine what is causing a problem as well. What this method does is force you to keep asking the question why, at least five times, so that you can really get to the root cause of the problem. It is like putting together the pieces of a jigsaw puzzle and you have to figure out what the missing piece is.

Once you have been able to go through and get the answers to these questions, it is time to make the right corrections. Even if you find that all you need is a simple change, make sure that you communicate and coordinate the change with everyone who will be affected. Remember that in Kanban, there isn't really a thing as overcommunication.

Any time that you see there is a problem occurring, you should also go back to the sizing calculations that you originally start with. You can review your assumptions and when necessary, you can recalculate the Kanban quantities using the changes. Once you have these figures, you can coordinate the changes and implement the new quantities that you want to use.

Auditing the Kanban system is very important for making sure that the system is as effective as possible. There are times when a part of the system may stop working. There may be a time when the requirements change and the system needs to change with these as well. Making minor changes on occasion to the Kanban system is not too difficult. The problems occur when you don't do an audit and these issues are left unchecked for a longer period of time. These relatively small issues can grow into big ones if they aren't fixed right away. The auditing process for Kanban is designed as a way to help deal with this process and try to keep everything on track.

Reviewing the Roles and Responsibilities for Kanban Implementation

As you work on maintaining your Kanban, it is important to take some time to review the different roles and responsibilities when it comes to implementing the Kanban system. The team needs to all get together and discuss the whole process. Not only should there be a discussion on which cards are used for every part of the program, there should be a discussion about how the board works and then everyone should be in agreement on what needs to happen during each stage of the process.

Implementing the Kanban system has to be a team effort. It can't be the idea of one person, and then that one person has to implement it all, maintain it, and make sure that the system is working. Everyone needs to have a role and a responsibility before the implementation even begins. This ensures that it can go off without a hitch and allows everyone to have some pride in the work that they do.

After the system is implemented, it is a good idea to review who is in charge of each part and make adjustments as needed. There may be times when you need to shift around the responsibilities to eliminate waste or make other changes to improve the amount of efficiency that is found in the process.

Kanban Management for Raw Materials

Kanban can be a great way to make sure that the raw materials that are needed for your product are always present, without having to keep a ton of inventory on hand and hoping that the customer will purchase that product. There are other ways to manage the raw materials for a product, but Kanban is becoming more popular because it is efficient, reduces the amount of waste in the process, and can save the business money.

If the Kanban system is set up the proper way, it will tell you when it is time to order more raw materials. A card will show up at the right time, alerting the proper team or the proper person that it is time to make the order. Now, this will take a bit of time and adjusting in the beginning. Someone has to go in and determine the right place to put the card so that the raw material will come in time to keep making the product, without any delays.

This is where the idea of the lead time and the cycle time comes into play. If you know that it takes two days to get more raw materials in, then the Kanban card needs to be placed so that you can make it two days with your current supply until the new materials get in. So, if you go through 100 units of the product each day, and it takes two days to get new products in, the Kanban card should be placed when you get to 200 units. This way, you can keep making the product and order more in, without running into a shortage.

You also need to make adjustments to this system on occasion. With the example above, you may need 100 units a day during regular times. But then during the holiday season, you end up going through 150 units each day. You would need to place the card in a different location. Otherwise, you will run into a shortage of the product and customers who are not happy waiting for more product.

This can go the other way as well. You may find that you end up with too much inventory because your average for the raw material switches down to 80 units a day, instead of the original 100 units a day. When this happens, you may need to move the Kanban cards to adjust for this, ensuring that you still get the product that you want, but also making sure that you don't get too much product in at a time.

Knowing your lead time, and knowing how much is used during a cycle, is critical when it comes to working with the Kanban system. It allows you to keep your inventory to a minimum, while still providing the product quickly and efficiently to the customer. You will have to go through and make adjustments to the system on occasion to ensure that the cycle time is still the same. If it has changed, then you need to change things up as well. And if the lead time does change, such as taking three days instead of two, you will need to go through and make those adjustments as well to ensure you have enough products for your demand.

Six Rules to Make the Kanban System Effective

To make sure that you are setting up the Kanban system properly in your workplace, you need to follow the six rules that Toyota has provided to make the system effective. These six rules include:

1. The downstream, or the customer, is going to process the withdraw items in the precise amounts that are specified by the Kanban.

2. The upstream, or the supplier, will produce the items, only in the exact amounts and sequences specified by the Kanban system.

3. No items are made or moved without a Kanban.

4. The Kanban needs to accompany each of the items that you did every time.

5. Defects and the amounts that are incorrect are never sent to the next downstream process.

6. The number of Kanbans will be reduced at a careful manner in order to lower inventories and then reveals problems.

Conclusion

Thank you for making it through to the end of *Kanban*. Let's hope it was informative and able to provide you with all of the tools you need to achieve your goals whatever they may be.

The next step is to implement some of the steps that we discussed in this guidebook in order to implement a Kanban system in your own workplace. There are a lot of different systems out there to help with the production process, no matter what kind of product you are trying to sell. But none of them seem to be as good at reducing waste and inefficiencies as the Lean methodology, and Kanban is a visual way to implement those ideas into your team and get all of the benefits at the same time.

This guidebook spent some time looking at different parts that come with the Kanban system and how to implement it into your own company. Even though we discussed some of the basics in the other two books, this guidebook has started out with a short introduction to Kanban to help us review before getting into more of the information.

After we take a look at the Kanban system and some of the things that Kanban can do, we have moved on to learning more about the definitions and words and language that are involved in the Lean environment and that have been useful as we progress through this guidebook. We then looked at some of the different diagrams that have been used successfully in the Kanban process, how to prepare for Kanban to get the most benefits, and then we took a look at the importance of a good supply chain management and how the process of Kanban can make that easier.

From there, we took a look at the ideas of ABC classification and how this was a useful tool to help you split up your products and your customers in order to cater to each one in the most efficient way. We even spent some time talking about the traditional MRP system that has been used in many companies for years and how it compares to the Kanban system that many companies are adopting now.

To end this guidebook, we looked at some of the steps that were needed in order to design an efficient Kanban system to get the most out of it, how to implement that system, and some of the steps that are needed to maintain your Kanban system. The best system in the world is going to be worthless if no one follows or maintain it, so make sure that you spend some extra time following the ideas in that chapter.

The Lean methodology, with the help of the Kanban system that we discussed in-depth in this guidebook, has done a lot of good in the way that many businesses run their own production line. Make sure that you have to read through this guidebook to help you get started in the right direction.

Finally, if you found this book useful in any way, a review on Amazon is always appreciated!

CPSIA information can be obtained
at www.ICGtesting.com
Printed in the USA
LVHW032234050319
609644LV00001B/125/P